CAMBRIDGE TRACTS IN MATHEMATICS

General Editors

B. BOLLOBAS, H. HALBERSTAM & C.T.C. WALL

---

90 *Divisors*

T0296411

RICHARD R. HALL

Reader in Mathematics,
University of York

GÉRALD TENENBAUM

Chargé de Recherches au CNRS,
Département de Mathématiques de l'Université de Nancy I

Divisors

The right of the University of Cambridge to print and sell all manner of books was granted by Henry VIII in 1534. The University has printed and published continuously since 1584.

CAMBRIDGE UNIVERSITY PRESS

Cambridge

New York New Rochelle Melbourne Sydney

CAMBRIDGE UNIVERSITY PRESS
Cambridge, New York, Melbourne, Madrid, Cape Town, Singapore, São Paulo, Delhi

Cambridge University Press
The Edinburgh Building, Cambridge CB2 8RU, UK

Published in the United States of America by Cambridge University Press, New York

www.cambridge.org
Information on this title: www.cambridge.org/9780521340564

First published 1988
This digitally printed version 2008

A catalogue record for this publication is available from the British Library

Library of Congress Cataloguing in Publication data

Hall, R.R. (Richard Roxby)
Divisors.
(Cambridge tracts in mathematics; 90)
Bibliography: p.
1. Divisor theory. 2. Distribution (Probability
theory) I. Tenenbaum, G. (Gérald) II. Title.
III. Series.
QA242.H29 1988 519.2 87-24239

ISBN 978-0-521-34056-4 hardback
ISBN 978-0-521-09167-1 paperback

to Barbara,

à Catherine,

Contents

Preface

We have written this book in the belief that the study of the divisors of integers can be pursued much further than is generally accepted, not only to widen our understanding of the multiplicative structure of the integers but also to yield results of value in other branches of the theory of numbers.

It is too early to try to be at all definitive because there are fundamental problems about divisors which have not been resolved. However, it is possible to offer a structured description of at least part of the theory, in which the results follow each other in a comprehensible, natural fashion toward conclusions which seem inevitable and part of a distinct discipline.

Each positive integer n has $\tau(n)$ positive divisors including itself and unity, which we may order in increasing sequence as $\{d_i(n): 1 \leqslant i \leqslant \tau(n)\}$. We call this the *additive ordering*, induced on the set of divisors of n by the order on $\mathbb{Z}$ generated by the addition law. Besides this ordering (obtained naturally as the trace of the ordering on a bigger set) there is another one, arising from the decomposition into prime factors, which we call the *multiplicative ordering*. Let $n = p_1^{\alpha_1} p_2^{\alpha_2} \dots p_r^{\alpha_r}$ where $p_1 < p_2 < \cdots < p_r$; then each divisor d may be written in the form $p_1^{\beta_1} p_2^{\beta_2} \cdots p_r^{\beta_r}$ with $0 \leqslant \beta_i \leqslant \alpha_i$ $(1 \leqslant i \leqslant r)$. The new order is just the lexicographical one on the words $\beta_r \beta_{r-1} \dots \beta_1$. (A similar ordering of the sequence of squarefree numbers, which then begins 1, 2, 3, 6, 5, 10, 15, 30, 7, ... is possible, but not of $\mathbb{Z}^+$ itself.) Many of the interesting problems about divisors arise from the conflict between the two orders – we mention just one here. For any squarefree n with r prime factors, the sequence $\{\mu(d): d|n\}$ with the divisors in multiplicative order coincides with the first 2^r terms of the famous Thue–Morse sequence starting 1, – 1, – 1, 1, – 1, 1, 1, – 1, – 1, ..., and if we define

$$M(n) := \max_z \left| \sum_{\substack{d|n \\ d \leqslant z}} \mu(d) \right| \quad (n \in \mathbb{Z}^+)$$

then $M(n) \equiv 1$. However, in the additive ordering $M(n)$ presents serious difficulties, and the conjecture of Erdős and Hall (1980) that $M(n) \geqslant 2$ p.p. (we use 'presque partout' to mean 'on a sequence of asymptotic density 1') has only recently been settled by H. Maier.

The modern theory of divisors was initiated by Hardy & Ramanujan (1917) who proved that, for every fixed $\varepsilon > 0$,

$$(\log n)^{\log 2 - \varepsilon} < \tau(n) < (\log n)^{\log 2 + \varepsilon} \quad \text{p.p.,} \tag{1}$$

and defined the *normal order* of an arithmetical function. (In the text we refer to (1) as the Hardy–Ramanujan Theorem.) This is a theorem about the multiplicative structure of a large 'random' or normal integer, and immediately further questions about the distribution of the prime factors and divisors of such integers arise. One development from this paper was the Turán–Kubilius inequality and the theory of additive functions surrounding it – in fact we hardly need this, replacing it by the general principle explained in §0.5 below. What we have called the parametric method is almost embarrassingly simple in principle and often very precise.

The person who really understood Hardy and Ramanujan's paper was Erdős. Erdős saw that very precise statements about prime factors and divisors could be made p.p. but – and this is where much of the interest and difficulty of the subject lie – in many cases one would have to develop a machinery to tackle the divisors directly. The information available from best possible p.p. statements about prime factors would not be sufficient to settle the more delicate questions about divisors. In Chapter 1 we describe the most important results about the distribution of the prime factors which are needed later in the book, particularly the law of the iterated logarithm, and we indicate their limitations in the study of divisors.

Erdős' early interest in abundant numbers (perhaps not very important themselves) led him to consider the much more important ideas involved in *sets of multiples* and *primitive sequences*. He was one of the principal founders of the theory of sets of multiples, beautifully set out in Chapter V of Halberstam and Roth's *Sequences*. We shall not treat sets of multiples directly in the present book – the subject will some day require a book to itself – and for the same reason we have also avoided the closely related topic of divisor density. However, it was realized quite soon by Besicovitch that to study sets of multiples one needs information about divisors in 'short' intervals; this falls within the scheme of the book and is treated in Chapter 2.

Since Hilbert's solution of Waring's Problem at the beginning of the century, and Hardy's creation of the circle method which followed, it is the additive theory which has dominated analytic number theory: not so much the theory of bases of Schnirelman, Mann and others but the great edifice built upon exponential sums by Weyl, Vinogradov and their followers and enriched by the input from algebraic geometry. There is much work to be done on multiplicative problems and the possibility of useful cross-fertilization. A very perceptive step in this direction was taken by Hooley in his paper 'On a new technique and its applications to the theory of numbers' (1979) and it is striking that exactly the same sort of problems about divisors arise here as with sets of multiples.

This has influenced our decisions about the content of the book. We believe that a mathematical discipline must have a comprehensible structure in which the theorems appear in a logical order as ineluctable and final. Most mathematicians will be more prepared to master difficult technical proofs if they can see a natural progression of ideas and an ultimate way forward. We have therefore restricted our field of view to the closely related areas of short intervals and the propinquity of divisors, on the one hand because of the applications through Hooley's technique and on the other because of the structure which can be perceived.

The divisors of an integer n occupy an interval of logarithmic length $\log n$. This is the average order of $\tau(n)$, but from the Hardy–Ramanujan Theorem a normal, random integer will have only about $(\log n)^{\log 2}$ divisors. Hence there are long intervals without divisors and the questions arise whether there are short intervals with more than one, or indeed many, divisors, and just how close on a logarithmic scale the closest divisors are. In Chapter 4 we introduce the most useful measures of the propinquity of divisors which have been formulated by various authors, and relate them where possible. The emphasis is on the problem of whether there are relatively large numbers of divisors in intervals of logarithmic length $(\log n)^{\alpha}$, where $0 < \alpha < 1$, although the case $\alpha = 0$ is also considered.

Chapter 5 concentrates on the case $\alpha \leqslant 0$, that is really short intervals. Here the central problem from an historical point of view is Erdős' conjecture, nearly half a century old, that

$$\min\{d'/d: d, d'|n, d < d'\} \leqslant 2 \quad \text{p.p.},$$

recently confirmed by Maier and Tenenbaum. In fact the 2 on the right may be replaced by $1 + (\log n)^{-\kappa}$ for small positive κ – the supremum of the admissible κ is determined.

In these studies the Fourier transform is an invaluable tool, and this means that we need information about the function

$$\tau(n, \theta) := \sum_{d|n} d^{i\theta} \quad (= \sigma_{i\theta}(n))$$

and its generalizations. This part of the theory is set out in Chapter 3. The final two chapters are devoted to a particular measure of propinquity, Hooley's function

$$\Delta(n) := \max_{u} \operatorname{card}\{d: d|n, u < \log d \leqslant u + 1\}$$

and its generalization $\Delta_r(n)$. We are concerned with average orders. This topic fits naturally into the theory, and has the advantage for the student that the way to apply the results to, say, Waring's Problem has been worked out by Hooley, so that there is an immediately tangible objective.

The book is addressed to graduates training in analytic number theory, and to more advanced persons to whom the content appears unfamiliar and potentially useful.

We should like to thank Heini Halberstam for his encouragement, and Bob Vaughan for reading the finished manuscript. We are indebted to Michel Balazard, who worked through the earlier chapters in fine detail.

Notation

$(x)^+$	$= \max(x, 0)$
$\delta(x, y)$	$= 1$ if $x = y$, 0 else
$\log_k x$	$= \log \ldots \log x$ (k-fold iterated logarithm)
$p_i(n)$	the ith prime factor of the number n (in increasing sequence)
$d_i(n)$	the ith (positive) divisor of the number n (in increasing sequence)
$P^+(n)$, $P^-(n)$	the greatest, and least, prime factor of n (by convention $P^+(1) = 0$, $P^-(1) = \infty$)
$\omega(n)$	the number of distinct prime factors of n
$\omega(n, t)$	the number of distinct prime factors p of n such that $p \leqslant t$
$\omega(n, s, t)$	the number of distinct prime factors p of n such that $s < p \leqslant t$
$\Omega(n)$, $\Omega(n, t)$, $\Omega(n, s, t)$	similar to ω, but the prime factors are counted according to multiplicity
$\Omega(n, E)$	the number of prime factors, counted according to multiplicity, in a set E
$\tau(n)$	the number of (positive) divisors of n
$\tau_r(n)$	the number of ways of writing n as a product of r positive factors (different orders counted multiply)
$\tau(n, y, z)$	the number of divisors d of n such that $y < d \leqslant z$
$\mathscr{B}(\mathscr{A})$	the set of multiples of the sequence $\mathscr{A}$
$\mathbf{d}(\mathscr{A})$	the asymptotic density of $\mathscr{A}$
$\bar{\mathbf{d}}(\mathscr{A})$, $\underline{\mathbf{d}}(\mathscr{A})$	the upper, and lower, asymptotic density of the sequence $\mathscr{A}$
$\boldsymbol{\delta}(\mathscr{A})$	the logarithmic density of $\mathscr{A}$
p.p.	on a sequence of asymptotic density 1
p.p.x.	for $x + o(x)$ integers not exceeding x

The following is an index, by section numbers, of notation introduced in the text.

$H(y)$	0.2
$\Psi(x, y)$, $\Phi(x, z)$, $\Theta(x, y, z)$	0.3
$Q(\lambda) = \lambda \log \lambda - \lambda + 1$ $(\lambda > 0)$	0.4
$H(x, y, z)$	2.1
ρ (Dickman's function)	2.3
$\tau(n, \theta)$	3.1
$\rho(n, \theta)$	3.3
$U(n)$	3.3
$\zeta(s, y)$	3.4
$\mathbf{D}(\mathscr{A})$ (the divisor density of $\mathscr{A}$)	Chapter 3, Notes
$T(n, \alpha)$, $U(n, \alpha)$	4.2
$\tau^+(n)$	4.6
$\Delta(n)$	4.7
$E(n)$	5.1
$\Delta(n; u_1, u_2, \ldots, u_{r-1})$	6.1
$\Delta_r(n)$	6.1
$\alpha(r, y)$	6.1
Λ_r^-, Λ_r^+	6.2
H_r	6.7
$\mathscr{C}(F)$	Appendix

0
Preliminaries

0.1 Introduction

In this chapter we collect together some results and general principles which we use at several points in the book. We give a proof when we feel this may be unfamiliar to the reader and no standard reference is available; occasionally we merely sketch a proof in the exercises.

0.2 Sums of multiplicative functions

In most instances the following simple results will be sufficient for our purposes.

Theorem 00 (Halberstam and Richert). *Let f be a non-negative multiplicative function such that, for some $\kappa > 0$,*

$$\sum_{p \leqslant y} f(p) \log p \leqslant \kappa y + O\left(\frac{y}{\log^2 y}\right) \quad (y \geqslant 2), \tag{0.1}$$

$$\sum_{p \geqslant y} \sum_{\nu \geqslant 2} \frac{f(p^\nu)}{p^\nu} \log p^\nu \ll \frac{1}{\log y} \quad (y \geqslant 2). \tag{0.2}$$

Then, for $x \geqslant 2$, we have

$$\sum_{n \leqslant x} f(n) \leqslant \frac{\kappa x}{\log x}\left(1 + O\left(\frac{1}{\log x}\right)\right) \sum_{n \leqslant x} \frac{f(n)}{n}, \tag{0.3}$$

where the O-constant depends at most on κ and the implied constants in (0.1) *and* (0.2).

Notice that we have, on the right,

$$\sum_{n \leqslant x} \frac{f(n)}{n} \leqslant \prod_{p \leqslant x}\left(1 + \frac{f(p)}{p} + \frac{f(p^2)}{p^2} + \cdots\right), \tag{0.4}$$

the form in which the result is often applied. The constant κ in (0.3) is sharp (see the Notes).

For many purposes a weakened form of the inequality without the sharp constant is adequate – and this permits the straightforward proof given in Exercise 00.

Theorem 01. *Let f be a non-negative multiplicative function such that, for some A and B,*

$$\sum_{p \leqslant y} f(p) \log p \leqslant Ay \quad (y \geqslant 0), \tag{0.5}$$

$$\sum_{p} \sum_{\nu \geqslant 2} \frac{f(p^\nu)}{p^\nu} \log p^\nu \leqslant B. \tag{0.6}$$

Then, for $x > 1$,

$$\sum_{n \leqslant x} f(n) \leqslant (A + B + 1) \frac{x}{\log x} \sum_{n \leqslant x} \frac{f(n)}{n}. \tag{0.7}$$

For (0.2) or (0.6) to hold (with $B = B(\lambda_1, \lambda_2)$), it is sufficient that the Wirsing condition

$$f(p^\nu) \leqslant \lambda_1 \lambda_2^{\nu-1} \quad (\nu \geqslant 2, \quad \lambda_1 \geqslant 0, \quad 2 > \lambda_2 \geqslant 0)$$

is satisfied.

Theorem 02. *Let $A > 0$, $0 < \alpha < 1$. Then there exists $y_0 = y_0(A, \alpha)$ so that for every (complex valued) multiplicative function f such that*

$$|f(p)| \leqslant B, \quad \sum_{p} \sum_{\nu \geqslant 2} |f(p^\nu)| p^{-\alpha\nu} \leqslant C, \tag{0.8}$$

and

$$f(p) = 1 \quad (p > y \geqslant y_0), \tag{0.9}$$

we have uniformly for $x \geqslant \exp\{A^{-1}(B+1) \log y \cdot \log_2 y\}$, $y \geqslant y_0(A, \alpha)$ that

$$\sum_{n \leqslant x} f(n) = P(f)x + O\left(x \exp\left(-A \frac{\log x}{\log y} \right) \right), \tag{0.10}$$

where

$$P(f) := \prod_p \left(1 - \frac{1}{p}\right)\left(1 + \frac{f(p)}{p} + \frac{f(p^2)}{p^2} + \cdots\right)$$

and the implied constant depends only on A, B, and C.

Proof. This is an exercise on Rankin's method (cf. Notes). Let g be the multiplicative function $f * \mu$, that is

$$g(n) = \sum_{d \mid n} \mu(d) f(n/d). \tag{0.11}$$

Then, for $\nu \geqslant 1$, $g(p^\nu) = f(p^\nu) - f(p^{\nu-1})$, so that $g(p) = 0$ for $p > y$ and

$$\sum_{\nu=0}^{\infty} g(p^\nu) p^{-\nu} = (1 - p^{-1}) \sum_{\nu=0}^{\infty} f(p^\nu) p^{-\nu}.$$

We invert (0.11), insert in the left-hand side of (0.10) and reverse the order of summation. Thus

$$\sum_{n\leqslant x} f(n) = \sum_{d\leqslant x} g(d)\left[\frac{x}{d}\right] = x\prod_p\left(1+\frac{g(p)}{p}+\frac{g(p^2)}{p^2}+\cdots\right)+\mathcal{R}$$

say, where

$$|\mathcal{R}| \leqslant \sum_{d\leqslant x}|g(d)| + \sum_{d>x}|g(d)|\frac{x}{d} \leqslant \sum_{d=1}^{\infty}|g(d)|\left(\frac{x}{d}\right)^{\beta}$$

for any $\beta\in[0,1]$. Hence

$$|\mathcal{R}| \leqslant x^{\beta}\exp\left\{\sum_{p\leqslant y}|g(p)|p^{-\beta}+\sum_p\sum_{\nu\geqslant 2}|g(p^{\nu})|p^{-\beta\nu}\right\}$$

$$\ll_{B,C} x^{\beta}\exp\left\{(B+1)\sum_{p\leqslant y}\left(\frac{1}{p}+(1-\beta)\frac{\log p}{p^{\beta}}\right)\right\}$$

by (0.8), provided $\beta\geqslant\alpha$. We choose $\beta = 1-2A/\log y$ ($\geqslant \alpha$ if $y\geqslant y_0(A,\alpha)$) so that

$$|\mathcal{R}| \ll_{A,B,C} x^{\beta}(\log y)^{B+1} \ll_{A,B,C} x\exp\left(-A\frac{\log x}{\log y}\right)$$

for the values of x given. This gives the result stated.

We need to be able to estimate from above short sums of multiplicative functions. We quote the following result of Shiu (1980).

Theorem 03. *Let f be a non-negative, multiplicative function such that both*

$$f(p^{\nu})\leqslant A^{\nu}\quad (p \text{ prime}, \nu\in\mathbb{Z}^{+}),\ f(n)\ll_{\varepsilon} n^{\varepsilon}$$

for some fixed A and every $\varepsilon>0$ respectively. Let $\alpha,\beta\in(0,\frac{1}{2})$ be fixed, $k<y^{1-\alpha}$, $x^{\beta}<y\leqslant x$, $(a,k)=1$. Then we have, uniformly in a, k, y and x, that

$$\sum_{\substack{x-y<n\leqslant x\\ n\equiv a \pmod k}} f(n) \ll_{f,\alpha,\beta} \frac{y}{\varphi(k)}(\log x)^{-1}\exp\left(\sum_{\substack{p\leqslant x\\ p\nmid k}} f(p)/p\right).$$

Furthermore, if we restrict to the class $\{f: f(n)\leqslant 1\}$ we may replace $\ll_{f,\alpha,\beta}$ by $\ll_{\alpha,\beta}$.

The last statement is important in our applications and does not appear in Shiu's paper; however it is implicit in his proof.

The sum

$$\sum_{u<n\leqslant v} h(n)y^{\Omega(n)}\quad (h\in C^1[u,v], 0<y<2)$$

also arises, and with this in mind we state a familiar result in a form which lends itself to partial summation.

Theorem 04. *Let $N \in \mathbb{Z}^+$, $y \in \mathbb{C}$, $0 < |y| < 2$. There exists a polynomial $F_{N,y}$ of degree N such that, for $x \geqslant 1$,*

$$\sum_{n \leqslant x} y^{\Omega(n)} = \int_1^x F_{N,y}\left(\frac{1}{\log 2t}\right) \log^{y-1} 2t \, dt$$
$$+ O_{N,y}(x \log^{\operatorname{Re} y - N - 2} 2x).$$

We have $F_{N,y}(0) = H(y)/\Gamma(y)$ where

$$H(y) := \prod_p \left(1 - \frac{y}{p}\right)^{-1} \left(1 - \frac{1}{p}\right)^y.$$

0.3 Sieve estimates

These concern the functions

$$\Psi(x, y) := \operatorname{card}\{n : n \leqslant x, P^+(n) \leqslant y\}, \tag{0.12}$$
$$\Phi(x, z) := \operatorname{card}\{n : n \leqslant x, P^-(n) > z\}, \tag{0.13}$$
$$\Theta(x, y, z) := \operatorname{card}\left\{n : n \leqslant x, \prod_{\substack{p^\nu \| n \\ p \leqslant y}} p^\nu > z\right\}, \tag{0.14}$$

for which we give simple uniform bounds. In each case, more precise results are known – we give references in the Notes. Sketch proofs are to be found in the Exercises.

Theorem 05. *Uniformly for $x \geqslant y \geqslant 2$,*

$$\Psi(x, y) \ll x \exp\left(-c_0 \frac{\log x}{\log y}\right). \tag{0.15}$$

Theorem 06. *Uniformly for $\frac{1}{2}x \geqslant z \geqslant 2$,*

$$\Phi(x, z) \asymp x/\log z. \tag{0.16}$$

Theorem 07. *Uniformly for $x \geqslant z \geqslant y \geqslant 2$,*

$$\Theta(x, y, z) \ll x \exp\left(-c_0 \frac{\log z}{\log y}\right). \tag{0.17}$$

0.4 Local distributions of prime factors and Poisson variables

Landau proved in 1900 that, for each fixed k,

$$\operatorname{card}\{n \leqslant x : \omega(n) = k\} \sim \frac{x}{\log x} \frac{(\log_2 x)^{k-1}}{(k-1)!} \quad (x \to \infty). \tag{0.18}$$

This result has been extended and generalized by many authors, and it remains true, whether $\omega(n)$ be replaced by the number of prime factors belonging to some subset of the primes or k is allowed to tend to infinity with x suitably slowly, that the local distribution is approximately Poisson – cf. Selberg (1954), Norton (1976, 1979, 1982), Tenenbaum (1980).

We require uniform upper bounds similar to those of Hardy & Ramanujan (1917). In particular, we state the following necessary generalization of a result of Halász (1972).

Theorem 08. *Let E be an arbitrary set of primes,* $p_0 = \min\{p: p \in E\}$, *and define*

$$E(x) = \sum_{\substack{p \leqslant x \\ p \in E}} \frac{1}{p}, \quad \Omega(n, E) = \sum_{\substack{p^\alpha \| n \\ p \in E}} \alpha. \tag{0.19}$$

Then, uniformly for $0 \leqslant k \leqslant (p_0 - \varepsilon) E(x)$, *we have both*

$$\operatorname{card}\{n: n \leqslant x, \Omega(n, E) = k\} \ll_{\varepsilon, p_0} x \mathrm{e}^{-E(x)} E(x)^k / k!$$

and

$$\sum \{n^{-1}: n \leqslant x, \Omega(n, E) = k\} \ll_{\varepsilon, p_0} (\log x) \mathrm{e}^{-E(x)} E(x)^k / k!$$

Proof. Let N_k denote the cardinality in question, and define

$$M_k := \sum \{m^{-1}: P^+(m) \leqslant x, \Omega(m, E) = k\}.$$

Then

$$N_k \log x = \sum_{\substack{n \leqslant x \\ \Omega(n,E) = k}} \sum_{p | n} \log p + \sum_{\substack{n \leqslant x \\ \Omega(n,E) = k}} \left(\log x - \sum_{p | n} \log p \right) = S_1 + S_2$$

say. We have

$$S_1 = \sum_{p \leqslant x} \log p \sum_{\substack{m \leqslant x/p \\ \Omega(mp,E) = k}} 1 \leqslant \sum_{\substack{m \leqslant x \\ k-1 \leqslant \Omega(m,E) \leqslant k}} \sum_{p \leqslant x/m} \log p \ll x(M_{k-1} + M_k).$$

For complex v, $|v| \leqslant p_0 - \varepsilon$, we have

$$\sum_{P^+(m) \leqslant x} \frac{v^{\Omega(m,E)}}{m} = \prod_{\substack{p \leqslant x \\ p \in E}} \left(1 - \frac{v}{p}\right)^{-1} \prod_{\substack{p \leqslant x \\ p \notin E}} \left(1 - \frac{1}{p}\right)^{-1} \ll_{\varepsilon, p_0} \mathrm{e}^{(\operatorname{Re} v - 1) E(x)} \log x.$$

We put $v = r\mathrm{e}^{i\theta}$ with $r = k/E(x)$. We have

$$M_k = r^{-k} \frac{1}{2\pi} \int_0^{2\pi} \mathrm{e}^{-ik\theta} \sum_{P^+(m) \leqslant x} \frac{v^{\Omega(m,E)}}{m} \, \mathrm{d}\theta$$

$$\ll_{\varepsilon, p_0} \log x \cdot \mathrm{e}^{-E(x)} E(x)^k k^{-k} \int_0^{2\pi} \mathrm{e}^{k \cos \theta} \, \mathrm{d}\theta.$$

It is a familiar fact that the integral on the right is $\ll k^{-1/2} \mathrm{e}^k$, and it follows from Stirling's formula that, for $k \geqslant 1$,

$$M_k \ll_{\varepsilon, p_0} \log x \cdot e^{-E(x)} E(x)^k/k!$$

and of course this holds for $k=0$ too. This proves the second part of the lemma, because the sum in question is $\leqslant M_k$. We also have

$$S_1 \ll_{\varepsilon, p_0} x \log x \cdot e^{-E(x)} \frac{E(x)^k}{k!}\left(1+\frac{k}{E(x)}\right),$$

and the last factor on the right may be deleted because $k \leqslant (p_0-\varepsilon)E(x)$. Next, from Hall (1974) we have, for every $h \in \mathbb{Z}^+$,

$$\sum_{n \leqslant x}\left(\log x - \sum_{p \mid n} \log p\right)^h \ll_h x.$$

Hölder's inequality therefore yields

$$S_2 \ll_h N_k^{1-1/h} x^{1/h} = N_k (x/N_k)^{1/h}.$$

We may assume, for a suitable $A = A(p_0)$, that $N_k > x(\log x)^{-A}$, else the required upper bound is trivial. We choose $h \geqslant 2A$ to obtain

$$N_k \ll_{\varepsilon, p_0} (\log x)^{-1} S_1 + (\log x)^{-1/2} N_k,$$

and the result follows. A further result of this sort has recently been obtained by Balazard.

To apply Theorem 08 we need upper estimates for the Poisson distribution and we quote the following result without proof (cf. Norton (1976)).

Theorem 09. *Let $x>0$, $0<\alpha<1<\beta$. Then*

$$\sum_{k \leqslant \alpha x} e^{-x} x^k/k! < \frac{1}{1-\alpha} \frac{e^{-Q(\alpha)x}}{\sqrt{(\alpha x)}}, \tag{0.20}$$

$$\sum_{k \geqslant \beta x} e^{-x} x^k/k! < \frac{1}{\beta - 1}\left(\frac{\beta}{2\pi x}\right)^{1/2} e^{-Q(\beta)x} \tag{0.21}$$

where

$$Q(\lambda) := \lambda \log \lambda - \lambda + 1. \tag{0.22}$$

0.5 A general principle

Let $f(y) = \sum_{n=0}^{\infty} a_n y^n$ where $a_n \geqslant 0$ for all n. Evidently, for $N \in \mathbb{R}^+$,

$$\sum_{n \leqslant N} a_n \leqslant \inf_{0<y\leqslant 1} y^{-N} f(y), \tag{0.23}$$

$$\sum_{n \geqslant N} a_n \leqslant \inf_{y \geqslant 1} y^{-N} f(y), \tag{0.24}$$

and, for $N \in \mathbb{Z}^+$,

$$a_N \leqslant \inf_{y>0} y^{-N} f(y). \tag{0.25}$$

For example, if $f(y) = e^y$, (0.25) gives $1/N! \leqslant (e/N)^N$, which is too large only by a factor $\sim \sqrt{(2\pi N)}$.

This rather trivial device can be very useful: in many circumstances it yields almost sharp bounds with little effort. Thus Theorem 01 gives, for $y < p_0(E)$,

$$\sum_{n \leqslant x} y^{\Omega(n,E)} \ll \frac{x}{\log x} \prod_{\substack{p \leqslant x \\ p \in E}} \left(1 - \frac{y}{p}\right)^{-1} \prod_{\substack{p \leqslant x \\ p \notin E}} \left(1 - \frac{1}{p}\right)^{-1} \ll x e^{(y-1)E(x)},$$

and, choosing $y = k/E(x)$ we almost obtain Theorem 08, falling short by a factor $\sqrt{k}$. This can be retrieved by making y complex and applying Cauchy's theorem (and the $k^{-1/2}$ may be viewed as arising from stationary phase): however, the strength of the principle is its applicability when the more sophisticated analysis fails, and in this way it underpins many of our proofs. The following application concerns the normal behaviour of the additive function

$$\omega(n, t) = \operatorname{card}\{p : p|n, p \leqslant t\}.$$

Theorem 010. *Let $\psi(t) \to \infty$ as $t \to \infty$, $\psi(t) \ll (\log_2 t)^{1/6}$. Then for $3 \leqslant t \leqslant x$ we have*

$$|\omega(n, t) - \log_2 t| < \psi(t)\sqrt{\log_2 t} \tag{0.26}$$

for all except $\ll x \exp(-\frac{1}{2}\psi(t)^2)$ of the integers $n \leqslant x$.

This is a generalization of the theorem of Hardy & Ramanujan (1917) on the normal order of $\omega(n)$. Moreover the number of exceptional integers is estimated fairly precisely: notice that the Turán–Kubilius inequality gives just $\ll x\psi(t)^{-2}$.

Proof. We set

$$f(y) = \sum_{n \leqslant x} y^{\omega(n,t)} \ll \frac{x}{\log x} \prod_{p \leqslant t} \left(1 + \frac{y}{p-1}\right) \prod_{t < p \leqslant x} \left(1 - \frac{1}{p}\right)^{-1} \ll_{y_0} x (\log t)^{y-1}$$

uniformly for $0 < y \leqslant y_0$, by Theorem 01. Let $\chi(n)$ denote the characteristic function of the set of integers n for which

$$|\omega(n, t) - \log_2 t| \geqslant \varepsilon \log_2 t.$$

We apply (0.23) with $N = (1 - \varepsilon)\log_2 t$ and $y = 1 - \varepsilon$, and (0.24) with

$N=(1+\varepsilon)\log_2 t$ and $y=1+\varepsilon$, to obtain

$$\sum_{n\leqslant x}\chi(n)\ll x(\log t)^{-Q(1-\varepsilon)}+x(\log t)^{-Q(1+\varepsilon)}$$

with Q as in (0.22). Since $Q(1+u)=\frac{1}{2}u^2+O(u^3)$ $(|u|\leqslant 1)$, we obtain our result on setting $\varepsilon=\psi(t)/\sqrt{\log_2 t}$.

An analysis of the above proof shows that it belongs to the same family as Rankin's method: one tries to mimic the characteristic function of a certain set of integers by a multiplicative function which is easy to sum. The choice of this multiplicative function is of course crucial and is usually performed in two steps: first the general shape is determined, allowing a free parameter (e.g. $(x/n)^\alpha$ in Rankin's method, $y^{\omega(n,t)}$ in the proof of Theorem 010), and secondly this parameter is optimized.

Thus the idea of the proof of Theorem 010 is that the sum

$$\sum_{n\leqslant x} y^{\omega(n,t)}$$

is dominated by the n for which $\omega(n,t)=(y+o(1))\log_2 t$. This had already been noticed by Turán in the case $t=x$ (see the Notes). The method is still effective in the study of large deviations.

Notes on Chapter 0

§0.2. The Halberstam–Richert theorem generalizes a result of Hall (1974). Although we are not yet able to take advantage of this in many of our applications, it should be noticed that the constant κ on the right of (0.3) is sharp. To see this, set

$$P(f,x) = \prod_{p \leqslant x} \left(1 - \frac{1}{p}\right)\left(1 + \frac{f(p)}{p} + \frac{f(p^2)}{p^2} + \cdots\right)$$

so that (0.3), (0.4) and Mertens' Theorem yield

$$\sum_{n \leqslant x} f(n) \leqslant \kappa e^{\gamma} x P(f,x)\left(1 + O\left(\frac{1}{\log x}\right)\right). \tag{1}$$

Set

$$f(p^{\alpha}) = \begin{cases} 0 & (p \leqslant x/\log x), \\ \kappa & (x/\log x < p \leqslant x). \end{cases}$$

The left-hand side of (1) is $\sim \kappa\pi(x)$, and $P(f,x) \sim e^{-\gamma}/\log x$.

A more precise form of the inequality (0.4) was found by Hildebrand (1984). This has the effect of multiplying the right-hand side of (1) by a factor $\leqslant 1$, depending on f. The more often $f(p) \geqslant 1$ for $p \geqslant x^{\varepsilon}$, the better the gain, but nothing more than a factor $(e^{-\gamma} + o(1))$ can be expected here: Hildebrand (1984a) stated that when $f: \mathbb{Z}^+ \to [0,1]$ it may be shown elementarily that

$$\frac{1}{\log x} \sum_{n \leqslant x} \frac{f(n)}{n} \geqslant P(f,x)\left(1 + O\left(\frac{1}{\log x}\right)\right).$$

By *Rankin's method*, we mean the device of majorizing the characteristic function of the set $\{n: n \leqslant x\}$ by the weight $(x/n)^{\beta}$ with a suitable $\beta > 0$ which may be optimized. This can be surprisingly effective, cf. Rankin (1938).

The usual form of the right-hand side in Theorem 04, obtained by contour integration, is (for $x \geqslant 2$),

$$\frac{H(y)}{\Gamma(y)} x \log^{y-1} x \cdot \left\{1 + \frac{c_1(y)}{\log x} + \cdots + \frac{c_N(y)}{\log^N x} + O\left(\frac{1}{\log^{N+1} x}\right)\right\}.$$

§0.3. For a recent study of $\Psi(x,y)$, in particular an estimate of the loss involved in Rankin's method, see Hildebrand & Tenenbaum (1986). This paper contains an extensive list of references to previous work on $\Psi(x,y)$.

The proof of Theorem 06 suggested in Exercise 02 is quite different from that of Halberstam & Roth (1966) (q.v.). A stronger form of Theorem 07 appears in Tenenbaum (1984). In particular we have

$$\Theta(x,y,z) < x \exp\left\{-u\left(\log u + \log_2 u - 1 + O\left(\frac{\log_2 u}{\log u}\right)\right)\right\}$$

with $u := \log z/\log y$, uniformly for $u \leqslant y^{1/2-\varepsilon}$.

§0.5. Turán's variance proof of the Hardy–Ramanujan theorem is now classical – see e.g. Elliott (1979) – and is regarded as the actual birth of Probabilistic Number Theory. In the same year, 1934, Turán noticed in his Doctoral Dissertation (published in Hungarian) that the theorem could also be proved analytically, starting with the asymptotic formula (uniform for suitable y),

$$\sum_{n \leqslant x} y^{\omega(n)} = (C(y) + o(1))x \log^{y-1} x, \tag{2}$$

obtained by contour integration. This is explained in detail by Elliott (1979, Part II, pp. 18–20) where an enlightening letter of Turán is quoted. As Turàn remarked, (2), extended to complex y, is closer to the Erdős–Kac Theorem than to the Hardy–Ramanujan Theorem. On the other hand the method set out in §0.5 requires just the O-form of (2). This explains the fairly wide range of applicability of the principle: the finer analytical tools which are not always available are replaced by simple and general inequalities such as that of Halberstam and Richert.

Exercises on Chapter 0

00. Let f satisfy the conditions in Theorem 01 and set $S(x) = \sum\{f(n): n \leqslant x\}$, $M(x) = \sum\{f(n)/n: n \leqslant x\}$. Prove that $S(x)\log x \leqslant S_1 + S_2 + S_3$ where

$$S_1 = \sum_{p \leqslant x} f(p)\log p \sum_{m \leqslant x/p} f(m),$$

$$S_2 = \sum_{\nu \geqslant 2}\sum_{p^\nu \leqslant x} f(p^\nu)\log p^\nu \sum_{m \leqslant x/p^\nu} f(m)$$

and $S_3 \leqslant xM(x)$. Prove that $S_1 \leqslant AxM(x)$, and, by first showing that $S(z) \leqslant zM(z)$, for any z, prove that $S_2 \leqslant BxM(x)$. Deduce the Theorem.

01. Let $y \geqslant 2$, $\varepsilon \ll 1/\log y$. Prove that

$$\sum_{p \leqslant y} \frac{(p^\varepsilon - 1)}{p} = O(1).$$

Let $\chi(n, y) = 1$ if $P^+(n) \leqslant y$, $= 0$ else. Show that, for $\varepsilon > 0$,

$$\Psi(x, y) \leqslant \sqrt{x} + \sum_{n \leqslant x}\left(\frac{n}{\sqrt{x}}\right)^\varepsilon \chi(n, y)$$

and apply Theorem 01 to the sum on the right, and choose ε suitably, to prove (0.15).

02. By means of Theorem 02, show that

$$\Phi(x, z) = x\prod_{p \leqslant z}\left(1 - \frac{1}{p}\right) + O(x/\log^2 z),$$

uniformly for $x \geqslant \exp(2\log z \cdot \log_2 z)$. Show that $\Phi(x, z) \ll x/\log z$ $(2 \leqslant z \leqslant x)$ by using Theorem 01, and that $\Phi(x, z) \geqslant \pi(x) - \pi(z) \gg x/\log z$ for $x^{1/4} < z \leqslant \frac{1}{2}x$ by the Prime Number Theorem.

Now let $2 \leqslant z \leqslant x^{1/4}$. By means of the inequality $\log x \geqslant \sum\{\log p : p \mid n\}$ $(1 \leqslant n \leqslant x)$, show that, for $x > x_0$,

$$\Phi(x, z)\log x \gg x \sum_{m \leqslant \sqrt{x}} \chi(m)/m$$

where χ is the multiplicative function such that $\chi(p) = 1$ $(z < p \leqslant x^{1/3})$, $\chi(p^\nu) = 0$ (else). Prove that

$$\sum_{m \leqslant \sqrt{x}} \frac{\chi(m)}{m} \geqslant \prod_{z < p \leqslant x^{1/3}}\left(1 + \frac{1}{p}\right) - 2\sum_{m=1}^{\infty} \frac{\chi(m)}{m}\frac{\log m}{\log x}.$$

By logarithmically differentiating $\prod(1 + p^{-s})$ at $s = 1$, or otherwise, prove that the sum on the right is

$$\leqslant \left(\frac{2}{3}+O\left(\frac{1}{\log x}\right)\right) \prod_{z<p\leqslant x^{1/3}} \left(1+\frac{1}{p}\right)$$

and hence complete the proof of Theorem 06.

03. Show that

$$\Theta(x,y,z) \leqslant \sum_{\substack{z<a\leqslant x/y \\ P^+(a)\leqslant y}} \Phi(x/a,y)+\Psi(x,y),$$

and deduce Theorem 07, using in turn Theorems 01, 05 and partial summation.

04. Show that, for $y\leqslant(\geqslant)1$ respectively,

$$\operatorname{card}\{n\leqslant x: \omega(n)\leqslant(\geqslant) y\log\log x\} \ll_{y_0} x(\log x)^{-Q(y)},$$

(with Q as in (0.22)) uniformly in any finite range $0<y\leqslant y_0$. Show that, provided $y_0<2$, these estimates hold with ω replaced by Ω.

05. By writing $y^{\Omega(n)}=\sum_{d|n} g(d)$, prove that, for $1\leqslant y<2$,

$$\sum_{n\leqslant x} y^{\Omega(n)} \leqslant x \prod_{p\leqslant x}\left(1+\frac{y-1}{p-y}\right) \ll \frac{x\log x}{2-y}.$$

By choosing $y=2-1/k$, prove that

$$\operatorname{card}\{n\leqslant x: \Omega(n)=k\} \ll xk\cdot 2^{-k}\log x,$$

uniformly for $k\geqslant 1$, $x\geqslant 2$. This improves on Erdős–Sárközy (1980); for an asymptotic formula see Nicolas (1984).

06. Let δ be fixed, $0<\delta<\frac{1}{2}$, and set

$$\mathscr{A}(\delta) := \{m: m\geqslant 3, \omega(m)\leqslant(\tfrac{1}{2}+\delta)\log_2 m\}.$$

Show that $\mathbf{d}\mathscr{A}(\delta)=0$. The remainder of this exercise is devoted to proving

$$\sum_{\substack{d|n \\ d\in\mathscr{A}(\delta)}} 1 \sim \tau(n) \quad \text{p.p.} \tag{1}$$

(a) Show that, for any fixed $\varepsilon\in(0,1)$,

$$\operatorname{card}\{p: p|n, \log_2 p>(1-\varepsilon)\log_2 n\} \sim \varepsilon\omega(n) \quad \text{p.p.},$$
$$\operatorname{card}\{d: d|n, \log_2 d>(1-\varepsilon)\log_2 n\} \sim \tau(n) \quad \text{p.p.}$$

(b) Establish the inequality

$$\sum_{d|n} y^{\omega(d)} \leqslant (1+y)^{\Omega(n)} \quad (y \geqslant 0, n \in \mathbb{Z}^+),$$

and use it to prove that, for any $\eta \in (0, \frac{1}{2})$,

$$\operatorname{card}\{d : d|n, \omega(d) \leqslant (\tfrac{1}{2} + \eta)\Omega(n)\} \sim \tau(n) \quad \text{p.p.}$$

Deduce (1).

[$\mathscr{A}(\delta)$ is an example of a sequence having asymptotic density 0 and *divisor density* 1 – see the Notes on Chapter 3 and Hall (1978) for the definition of divisor density, and for further developments Hall (1981), Tenenbaum (1982), Dupain, Hall & Tenenbaum (1982), Hall & Tenenbaum (1986)].

1
The normal distribution of the prime factors

1.1 The law of the iterated logarithm

The distribution of the prime factors of an integer determines that of the divisors: in order to study the latter in principle we only need to know the behaviour of the step-function

$$\Omega(n,t) := \sum_{\substack{p^\alpha \| n \\ p \leqslant t}} \alpha .$$

For the reason explained below, we ignore multiplicity and consider instead

$$\omega(n,t) := \sum_{\substack{p|n \\ p \leqslant t}} 1 .$$

If $\xi(n) \to \infty$ as $n \to \infty$, we have $\Omega(n,t) - \omega(n,t) \leqslant \Omega(n) - \omega(n) < \xi(n)$ p.p. (because $\sum p^{-2} < \infty$), and, within the approximations in the following theorems, the two functions are interchangeable. In this chapter we are concerned primarily with normal behaviour: our interest is in properties of integers belonging to a suitable sequence of asymptotic density 1.

Let $\psi(t) \to \infty$ as $t \to \infty$, $\psi(t) \ll (\log_2 t)^{1/6}$. We have seen in Theorem 010 that, for fixed t, $2 \leqslant t \leqslant x$, the inequality

$$|\omega(n,t) - \log_2 t| < \psi(t)\sqrt{\log_2 t} \tag{1.1}$$

holds for all except $\ll x \exp(-\frac{1}{2}(\psi(t))^2)$ of the integers $n \leqslant x$. Thus we essentially know p.p.x the pointwise behaviour of $\omega(n,t)$.

This is insufficient for our purpose: we need a uniform approximation for $\omega(n,t)$ as t varies if we are to describe the distribution of the divisors, and our aim now is to achieve an optimal result of this sort. We begin with an easy first attempt, just taking advantage of the explicit bound for the number of exceptions to (1.1).

Theorem 10. *Let $\xi(x) \to \infty$ as $x \to \infty$. For each fixed $\varepsilon > 0$ we have*

$$|\omega(n,t) - \log_2 t| < (1+\varepsilon)\sqrt{(2\log_2 t \cdot \log_3 t)} \quad (\xi(x) < t \leqslant x) \quad \text{p.p.}x. \tag{1.2}$$

We restate this as a result about the sequence $\{p_j(n): 1 \leqslant j \leqslant \omega(n)\}$ of distinct prime factors of n.

Corollary. *Let* $\xi(n) \to \infty$ *as* $n \to \infty$. *Then*

$$|\log_2 p_j(n) - j| < (1+\varepsilon)\sqrt{(2j \log j)} \quad (\xi(n) < j \leqslant \omega(n)) \quad \text{p.p.} \tag{1.3}$$

This is immediate from (1.2), taking $t = p_j(n)$.

Proof of Theorem 10. We remark that, since $\omega(n, t)$ is non-decreasing, it will be sufficient to prove that (1.2) holds, with ε replaced by $\varepsilon/2$, at a set of discrete *checkpoints* t_k, $t_1 < t_2 < \cdots$, provided $\log_2 t_{k+1} - \log_2 t_k < \varepsilon' \sqrt{(\log_2 t_k \cdot \log_3 t_k)}$ and $\varepsilon' = \varepsilon'(\varepsilon)$ is sufficiently small. We put $t_k = \exp\exp k$, and $\psi(t) = \sqrt{((2+2\varepsilon)\log_3 t)}$ in (1.1). This gives the desired inequality at all the checkpoints in the range $\xi(x) < t_k \leqslant x$, and a total of

$$\ll \sum_{k \geqslant \log_2 \xi(x)} x k^{-1-\varepsilon} = o(x)$$

exceptional integers $n \leqslant x$. This completes the proof.

Next, we envisage the best possible form of Theorem 10. Let χ_p denote the characteristic function of the multiples of p; then it is fundamental in Probabilistic Number Theory that the χ_p behave in many respects like independent random variables – this is the basis of the Kubilius Model (cf. Elliott (1979)). Now we have

$$\omega(n, t) = \sum_{p \leqslant t} \chi_p(n)$$

and the expectation of the right-hand side is $\sum\{1/p: p \leqslant t\} = \log_2 t + O(1)$; if the law of the iterated logarithm applied we should have

$$\limsup \Lambda(n, t) = +1, \liminf \Lambda(n, t) = -1 \quad \text{p.p.}, \tag{1.4}$$

where

$$\Lambda(n, t) := \frac{\omega(n, t) - \log_2 t}{\sqrt{(2 \log_2 t \cdot \log_4 t)}};$$

in particular the $\log_3 t$ in (1.2) would become $\log_4 t$.

Erdős (1946) announced (1.4) without proof. We supply the details here, since the matter is not as straightforward as might be assumed – see also the Notes. The general principle described in §0.5 will be employed systematically to ease the technical computations, playing effectively the rôle of the so called 'exponential bounds' which occur in probability theory.

Theorem 11. *Let* $\xi(x) \to \infty$ *as* $x \to \infty$. *Then*

$$|\Lambda(n, t)| < 1 + \varepsilon \quad (\xi(x) < t \leqslant x) \quad \text{p.p.}x. \tag{1.5}$$

Furthermore, if $\xi(x)$ increases sufficiently slowly, we have

$$\inf_{\xi(x)<t\leqslant x} \Lambda(n,t)\leqslant -1+\varepsilon, \quad \sup_{\xi(x)<t\leqslant x} \Lambda(n,t)\geqslant 1-\varepsilon \quad \text{p.p.}x. \tag{1.6}$$

Proof of Theorem 11. We begin with (1.5). First we observe that we may restrict our attention to the range $\xi(x)<t\leqslant x_1:=\exp\exp(\log_2 x-\xi(x))$, provided the growth of $\xi(x)$ is sufficiently slow. For, if $t>x_1$, we have

$$\begin{aligned}|\omega(n,t)-\log_2 t| &\leqslant \omega(n)-\omega(n,x_1)+|\omega(n)-\log_2 x|+\xi(x)\\ &<\xi(x)\sqrt{\log_2 x}\quad \text{p.p.}x;\end{aligned}$$

to see this, notice first that, since

$$\sum_{n\leqslant x}(\omega(n)-\omega(n,x_1))\leqslant \sum_{x_1<p\leqslant x} x/p \ll x\xi(x),$$

plainly $\omega(n)-\omega(n,x_1)<\frac{1}{3}\xi(x)\sqrt{\log_2 x}$ p.p.x – a similar bound for $|\omega(n)-\log_2 x|$ follows from the Hardy–Ramanujan Theorem. We shall therefore have

$$\sup_{x_1<t\leqslant x}|\Lambda(n,t)|=\mathrm{o}(1)\quad \text{p.p.}x$$

provided, and we assume this henceforth,

$$\xi(x)=\mathrm{o}(\sqrt{\log_4 x}). \tag{1.7}$$

Second, we note that we may also neglect *a priori* those integers $n\leqslant x$ which do not satisfy

$$\prod_{\substack{p^\alpha\|n\\ p\leqslant x_1}} p^\alpha\leqslant x^{1/4}. \tag{1.8}$$

Indeed by Theorem 07 the number of exceptional $n\leqslant x$ is $\mathrm{o}(x)$.

Now let S be the set of integers $n\leqslant x$ such that (1.8) holds. We are going to prove

$$\operatorname{card}\{n\in S:\ \sup_{\xi(x)<t\leqslant x_1}|\Lambda(n,t)|>1+\varepsilon\}=\mathrm{o}(x). \tag{1.9}$$

In order to modify the proof of Theorem 10 we need checkpoints T_j increasing so rapidly that the series $\sum\exp(-\frac{1}{2}\psi(T_j)^2)$ is convergent, with $\psi(T)$ essentially $(1+\varepsilon)\sqrt{(2\log_4 T)}$. However, we cannot then obtain a satisfactory bound for $|\omega(n,t)-\log_2 t|$ *between* the checkpoints merely from the monotonicity of $\omega(n,t)$, and another idea – involving some technical complication – is needed. Basically this is to use the old checkpoints *as well*, and to appeal to Theorem 01 to estimate $|\omega(n,t_k)-\omega(n,T_j)-(\log_2 t_k-\log_2 T_j)|$ when $t_k\in(T_j,T_{j+1}]$.

As before we put $t_k=\exp\exp k$ $(k\geqslant 1)$; it will be sufficient to prove (1.9) with t confined to the sequence $\{t_k\}$, and ε replaced by $\varepsilon/2$. For convenience we write $\varepsilon/2=\delta$. We set $I=[\delta^{-1}\log_3\xi(x)]$, $J=[\delta^{-1}\log_3 x_1]+1$, and, for each j, $I\leqslant j\leqslant J$, we write $K_j=\exp(\delta j)$ so that $K_I\leqslant\log_2\xi(x)$, $K_J>\log_2 x_1$. The new checkpoints are $T_j=\exp\exp K_j=\exp\exp\exp\delta j$ and for each of these j we denote by S_j the set of those $n\in S$ such that

$$\sup_{K_j<k\leqslant K_{j+1}}|\Lambda(n,t_k)|>1+\delta, \tag{S$_j$}$$

and by A_j the set of those $n\leqslant x$ for which

$$|\omega(n,T_{j+1})-\log_2 T_{j+1}|>\psi(T_j)\sqrt{\log_2 T_j}, \tag{A$_j$}$$

where we shall actually take

$$\psi(T)=(1+\delta)\sqrt{(2\log_4 T)}-c,$$

with a suitably large constant c (to be specified later). We show that

$$\sum_{I\leqslant j\leqslant J}\operatorname{card} S_j=o(x).$$

In fact we shall prove that, for $x>x_0(\delta)$, we have

$$\operatorname{card} S_j\leqslant 2\operatorname{card} A_j\quad(I\leqslant j\leqslant J), \tag{1.10}$$
$$\operatorname{card} A_j\ll_\delta xj^{-1-\delta/4}\quad(I\leqslant j\leqslant J), \tag{1.11}$$

thus establishing (1.9) and so (1.5).

The proof of (1.11) is easy. Condition (A_j) implies for $x>x_0(\delta)$ that

$$|\omega(n,T_{j+1})-\log_2 T_{j+1}|>(1+\tfrac{2}{3}\delta)\sqrt{(2\log_2 T_j\cdot\log_4 T_j)}$$

and, noticing that $\log_2 T_j=e^{-\delta}\log_2 T_{j+1}$, and substituting $\psi_1(T_{j+1})=(1+\tfrac{2}{3}\delta)\sqrt{(2e^{-\delta}\log_4 T_j)}$ in (1.1), we see that (1.11) follows.

To obtain (1.10), we split S_j into $K_{j+1}-K_j$ disjoint subsets $S_{j,k}$, $K_j<k\leqslant K_{j+1}$, defined by the conditions

$$\max_{K_j<m<k}|\Lambda(n,t_m)|\leqslant 1+\delta<|\Lambda(n,t_k)|. \tag{$S_{j,k}$}$$

Evidently

$$\operatorname{card} S_j\leqslant\sum_{K_j<k\leqslant K_{j+1}}\operatorname{card} S_{j,k}. \tag{1.12}$$

Let $B_{j,k}\subseteq S$ comprise the integers satisfying the extra condition

$$\begin{aligned}&|\omega(n,T_{j+1})-\omega(n,t_k)-(\log_2 T_{j+1}-\log_2 t_k)|\\&\leqslant c\sqrt{\log_2 T_j}\quad(K_j<k\leqslant K_{j+1}).\end{aligned} \tag{$B_{j,k}$}$$

Now the second condition $(S_{j,k})$ implies that

$$|\omega(n,t_k)-\log_2 t_k|>(1+\delta)\sqrt{(2\log_2 T_j\cdot\log_4 T_j)}$$

whence we see that any n belonging to $S_{j,k}\cap B_{j,k}$ satisfies (A_j), that is $S_{j,k}\cap B_{j,k}\subseteq A_j$. Since the sets $S_{j,k}$ (j fixed) are disjoint, we infer that

$$\sum_{K_j<k\leqslant K_{j+1}}\operatorname{card}(S_{j,k}\cap B_{j,k})\leqslant\operatorname{card} A_j, \tag{1.13}$$

so, in view of (1.12), (1.10) will follow from

$$\operatorname{card}(S_{j,k}\cap \bar{B}_{j,k})\leqslant \tfrac{1}{2}\operatorname{card} S_{j,k}, \tag{1.14}$$

where $\bar{B}_{j,k}=S\backslash B_{j,k}$. We shall see that (1.14) holds if c is sufficiently large.

Denote by a and b respectively generic integers for which $P^+(a)\leqslant t_k$, $P^-(b)>t_k$. We have

$$\operatorname{card} S_{j,k}=\sum_{\substack{a<x^{1/4}\\ a\in S_{j,k}}}\sum_{b\leqslant x/a} 1 \gg x\mathrm{e}^{-k}\sum_{\substack{a<x^{1/4}\\ a\in S_{j,k}}} a^{-1}$$

by Theorem 06. The hypothesis that (1.8) is satisfied is crucial at this point: it enables us to assume $a<x^{1/4}$ and hence $t_k=o(x/a)$ so that the sum over b above may be estimated by Theorem 06.

At this stage we notice that, for any y_1, y_2 satisfying $\frac{1}{2}\leqslant y_1<1<y_2\leqslant\frac{3}{2}$, we have

$$\operatorname{card}(S_{j,k}\cap\bar{B}_{j,k})\leqslant\sum_{\substack{a<x^{1/4}\\ a\in S_{j,k}}}\sum_{b\leqslant x/a}(y_1^{\alpha_1(b)}+y_2^{\alpha_2(b)})$$

where $\alpha_i(b):=\omega(b,T_{j+1})-(K_{j+1}-k)-\varepsilon_i c\sqrt{K_j}, \varepsilon_1=-1, \varepsilon_2=+1$. Indeed, for $n=ab$, we have $\omega(n,T_{j+1})-\omega(n,t_k)=\omega(b,T_{j+1})$, and if $n\notin B_{j,k}$ then either $\alpha_1(b)<0$ or $\alpha_2(b)>0$, whence $y_1^{\alpha_1}+y_2^{\alpha_2}>1$.

We apply Theorem 01. For $i=1,2$,

$$\sum_{b\leqslant x/a} y_i^{\alpha_i(b)}\ll\frac{x}{a}\mathrm{e}^{-k}\exp\{(K_{j+1}-k)(y_i-1-\log y_i)-c|\log y_i|\sqrt{K_j}\}$$

and we select $y_i=1+\varepsilon_i/2$ or $y_i=1+\varepsilon_i cK_j^{1/2}/(K_{j+1}-k)$ according as $k>K_{j+1}-2cK_j^{1/2}$ or not. A straightforward calculation shows that the expression in curly brackets in the exponential above is then $\leqslant-\beta c$, where β is an absolute constant, whence we have

$$\operatorname{card}(S_{j,k}\cap\bar{B}_{j,k})\ll x\mathrm{e}^{-k-\beta c}\sum_{\substack{a<x^{1/4}\\ a\in S_{j,k}}}\frac{1}{a}\ll\mathrm{e}^{-\beta c}\operatorname{card} S_{j,k}.$$

For suitable c, this yields (1.14), and hence (1.10).

Let us embark on the proof of (1.6). The required inequalities are symmetrical and we confine our attention to the second. We put $X:=\exp\{\log x/\log_2 x\}$ and we shall prove that

$$\sup_{\xi(x)<t<X}\Lambda(n,t)>1-\varepsilon\quad\text{p.p.}x. \tag{1.15}$$

Let $D=D(\varepsilon)$ be a large constant, to be specified later, and put $K=[\log_3\xi(x)/\log D]$, $L=[\log_3 X/\log D]$. For $K<k\leqslant L$, put $t_k=\exp\exp D^k$

and $I_k = (t_{k-1}, t_k]$, finally

$$\omega_k(n) := \text{card}\,\{p: p|n, p \in I_k\}.$$

The basic strategy now is to establish quasi-independence of the $\omega_k(n)$, sufficiently strong to permit us to follow the classical probabilistic approach. To this end, we estimate the characteristic function (in the sense of probability theory) of the vector

$$\boldsymbol{\omega}(n) := (\omega_{K+1}(n), \omega_{K+2}(n), \ldots, \omega_L(n))'.$$

We apply Theorem 02, with $A=3$, $B=2$ and $y=X$. This gives

$$\sum_{n \leqslant x} \prod_{K<k\leqslant L} z_k^{\omega_k(n)} = x \prod_{K<k\leqslant L} \prod_{p \in I_k} \left(1 + \frac{z_k - 1}{p}\right) + O(x/\log x) \tag{1.16}$$

uniformly for complex z_k, $|z_k| \leqslant 2$, $K < k \leqslant L$.

It will be convenient to employ the following terminology: if f and g are functions of many complex variables $z_{K+1}, z_{K+2}, \ldots, z_L$ defined by the power series

$$f(z_{K+1}, \ldots, z_L) = \sum a_{m_{K+1},\ldots,m_L} z_{K+1}^{m_{K+1}} \ldots z_L^{m_L}$$
$$g(z_{K+1}, \ldots, z_L) = \sum b_{m_{K+1},\ldots,m_L} z_{K+1}^{m_{K+1}} \ldots z_L^{m_L},$$

we say that g majorizes f and write $f \prec g$ if, for every $m_{K+1}, m_{K+2}, \ldots, m_L$,

$$|a_{m_{K+1},\ldots,m_L}| \leqslant b_{m_{K+1},\ldots,m_L}.$$

It is then almost trivial that $f_1 \prec g_1$ and $f_2 \prec g_2$ imply $f_1 f_2 \prec g_1 g_2$.

Because $e^{-1/p} > 1 - 1/p$ we have

$$1 + \frac{z-1}{p} \prec \exp\left(\frac{z-1}{p-1} + \frac{1}{p(p-1)}\right)$$

and hence

$$\prod_{K<k\leqslant L} \prod_{p \in I_k} \left(1 + \frac{z_k - 1}{p}\right) \prec \exp\left\{1 + \sum_{K<k\leqslant L} (z_k - 1) H_k\right\}$$

where

$$H_k := \sum_{p \in I_k} \frac{1}{p-1} \quad (K < k \leqslant L).$$

It follows that for vectors $\mathbf{j} \in (\mathbb{Z}^+)^{L-K}$ we have

$$N(x, \mathbf{j}) := \text{card}\,\{n: n \leqslant x, \boldsymbol{\omega}(n) = \mathbf{j}\} \leqslant ex \prod_{K<k\leqslant L} \left(e^{-H_k} \frac{H_k^{j_k}}{j_k!}\right) + \mathscr{R}$$

with

$$\mathscr{R} = (2\pi)^{K-L} \int_{C_{K+1}} \cdots \int_{C_L} O\left(\frac{x}{\log x}\right) \prod_{K<k\leqslant L} |dz_k| |z_k|^{-j_k - 1},$$

in which, for each k, C_k is the circle $|z_k| = r_k, 0 \leqslant r_k \leqslant 2$. We allow $r_k = 0$ only when $j_k = 0$: if C has radius 0 it is to be understood that

$$\frac{1}{2\pi}\int_C |dz||z|^{-1} = 1.$$

Let $j_k \leqslant 2H_k$ for all k, $K < k \leqslant L$. We choose $r_k = j_k/H_k$ above to deduce that

$$\mathscr{R} \ll \frac{x}{\log x}\prod_{K<k\leqslant L}(H_k/j_k)^{j_k} \ll x \prod_{K<k\leqslant L} e^{-H_k}\frac{H_k^{j_k}}{j_k!}$$

(since $\sum H_k \leqslant \log_2 x + O(1)$), whence

$$N(x,\mathbf{j}) \ll x \prod_{K<k\leqslant L} e^{-H_k}\frac{H_k^{j_k}}{j_k!} \tag{1.17}$$

provided $j_k \leqslant 2H_k$ $(K < k \leqslant L)$. We set $h_k = H_k + \sqrt{(2H_k \log(k \log D))}\cdot$ $(K < k \leqslant L)$. Since $H_k = D^k - D^{k-1} + O(1)$, certainly $h_k \leqslant 2H_k$ for sufficiently large D. Let $M(x)$ be the number of integers $n \leqslant x$ such that $\omega_k(n) \leqslant h_k$ for every k, $K < k \leqslant L$. From (1.17),

$$M(x) \ll x \prod_{K<k\leqslant L}\left(e^{-H_k}\sum_{j\leqslant h_k} H_k^j/j!\right).$$

Uniformly for $H \geqslant 1$, $\theta \geqslant 1$, we have

$$\sum_{j>H+\sqrt{(2H\theta)}} e^{-H}H^j/j! \gg \theta^{-1/2}e^{-\theta};$$

this just requires Stirling's formula and a straightforward computation. Therefore for a suitable absolute constant c_0,

$$M(x) \ll x \prod_{K<k\leqslant L}\left(1 - \frac{c_0}{k\log D\cdot\sqrt{\log(k\log D)}}\right) = o(x),$$

whence we have

$$\sup_{K<k\leqslant L}(\omega_k(n) - h_k) > 0 \quad \text{p.p.x.} \tag{1.18}$$

Let n be an integer not exceeding x such that both (1.5) and (1.18) hold, and let $\omega_l(n) - h_l > 0$. By (1.5),

$$\omega(n, t_{l-1}) \geqslant D^{l-1} - (1+\varepsilon)\sqrt{(2D^{l-1}\log(l\log D))},$$

whence

$$\begin{aligned}\omega(n,t_l) &= \omega(n,t_{l-1}) + \omega_l(n)\\ &\geqslant D^{l-1} + H_l\\ &\qquad + (\sqrt{H_l} - (1+\varepsilon)\sqrt{D^{l-1}})\sqrt{(2\log(l\log D))}\\ &> D^l + (1-\varepsilon)\sqrt{(2D^l\log(l\log D))}\end{aligned}$$

provided $D = D(\varepsilon)$ is sufficiently large. This proves (1.15).

1.2 The normal size of $p_j(n)$ and $d_j(n)$

An immediate corollary of Theorem 11 is the following sharpening of (1.3).

Theorem 12. *Let* $\varepsilon > 0$, *and* $\xi(n) \to \infty$ *sufficiently slowly as* $n \to \infty$. *Then*

$$|\log_2 p_j(n) - j| \leqslant (1+\varepsilon)\sqrt{(2j\log_2 j)} \quad (\xi(n) < j \leqslant \omega(n)) \quad \text{p.p.}; \tag{1.19}$$

furthermore, if $1+\varepsilon$ *be replaced by* $1-\varepsilon$ *on the right, the set of integers for which* (1.19) *is satisfied will have zero asymptotic density.*

We have therefore reached a best possible result concerning uniform approximation to $\omega(n, t)$ by the elementary function $\log_2 t$; and Theorem 11, and variants of it, will have applications in later chapters, where it is indispensable as a global statement about the distribution of prime factors. Regarding local information about neighbouring prime factors, Theorem 12 must be treated with caution if it is to be interpreted as meaning that $p_j(n)$ is approximately $\exp\exp j$. For example, this might lead us to expect, for almost all n and $j > \xi(n)$, that

$$p_j(n) > \prod_{i<j} p_i(n), \tag{1.20}$$

and this is completely false. The point is important and we give two demonstrations.

First, following Erdős & Kátai (1971) we define

$$M(n,z) = \sum_{\substack{d|n \\ d \leqslant z}} \mu(d), \quad M(n) = \sup_z |M(n,z)|. \tag{1.21}$$

If (1.20) held for $j > \xi(n)$ then $M(n) \leqslant 2^{\xi(n)}$, hence we should have $M(n) \leqslant \xi_1(n)$ p.p., provided $\xi_1(n) \to \infty$. But H. Maier (1987) has shown recently that

$$M(n) > (\log_2 n)^{\kappa - \varepsilon} \quad \text{p.p.} \quad \left(\kappa := \frac{-\log 2}{\log(1 - 1/\log 3)}\right),$$

confirming a conjecture of Erdős & Hall (1980).

In fact the truth about (1.20) is as follows. For $n \in \mathbb{Z}^+$ (not necessarily squarefree) and $y > 1$, let $\gamma(y, n)$ be defined by the relation

$$\max\{d : d|n, P^+(d) < y\} = y^{\gamma(y,n)}.$$

Then there exists a continuous, strictly increasing $G: \mathbb{R}^+ \to [0, 1)$, $G(0) = 0$, $G(c) \to 1$ as $c \to \infty$, such that

$$\frac{1}{\log_2 n} \operatorname{card}\{j : \gamma(p_j(n), n) \leqslant c\} \to G(c) \quad \text{p.p.} \tag{1.22}$$

(Erdős (1966)). For a more precise result, see Bovey (1977). Bovey showed that

$$G(c) = \mathrm{e}^{-\gamma} \int_0^c \rho(u)\,du$$

where γ is Euler's constant and ρ is Dickman's function (defined by (2.17)). In particular, (1.20) holds, p.p., for $(G(1) + o(1))\omega(n)$ values of j – indeed the influence of multiple factors can be neglected here since, plainly, $\prod_{p^\nu \| n} p^{\nu-1} \leqslant \xi(n)$, p.p. Notice that this implies a real difference between the two orderings of the divisors mentioned in the introduction.

A consequence of Theorem 12 is the following result about the jth divisor $d_j(n)$. A sketch proof is given in Exercise 12.

Theorem 13. *Let $\varepsilon > 0$ and $\xi(n) \to \infty$ as $n \to \infty$. Then we have*

$$\left| \log_2 d_j(n) - \frac{\log j}{\log 2} \right| \leqslant (1+\varepsilon) \sqrt{\left(\frac{2}{\log 2} \log j \cdot \log_3 j \right)}$$

$$(\xi(n) < j \leqslant \tau(n)) \quad \text{p.p.}; \qquad (1.23)$$

moreover the right-hand side is best possible (in the same sense as in Theorem 12).

The right-hand side is much too large to cope with delicate questions about divisors in short intervals or the propinquity of divisors. Once again we have an approximate formula, in this case

$$d_j(n) \approx \exp(j^{1/\log 2}) \qquad (1.24)$$

which, taken too literally, will lead to falsehood. Thus, if we define, for $0 \leqslant u \leqslant 1$,

$$F_n(u) = \frac{1}{\tau(n)} \operatorname{card}\{d : d|n, d \leqslant n^u\},$$

we might expect, arguing roughly from (1.24) that

$$F_n(u) \doteqdot u^{\log 2} \quad \text{p.p.} \qquad (1.25)$$

with some meaningful quality of approximation. The reality is quite different: if u_1 and u_2 are freely chosen from (0, 1) then $F_n(u_1) = F_n(u_2)$ on a sequence of positive density. The following 'principle of uncertainty' (to be proved in §2.3) shows in a rather strong way that F_n has no normal order. (See §4.4).

Theorem 14. *Let $\mathscr{A}$ be a sequence of integers such that F_n converges weakly when $n \to \infty$ through $\mathscr{A}$. Then $\mathbf{d}\mathscr{A} = 0$.*

These results on $F_n(u)$ may be found in Tenenbaum (1980). As we might expect, $F_n(u)$ has an average order. Uniformly for $0 \leqslant u \leqslant 1$,

$$\frac{1}{x}\sum_{n\leqslant x} F_n(u) = \frac{2}{\pi}\arcsin\sqrt{u} + o(1) \quad (x \to \infty). \tag{1.26}$$

(Deshouillers, Dress & Tenenbaum (1979)). So (1.25) is not even correct on average.

We end this chapter with a final, but important example warning against incautious use of (1.24). Erdős conjectured that

$$\min_{1\leqslant j<\tau(n)} (d_{j+1}(n)/d_j(n)) \leqslant 2 \quad \text{p.p.}, \tag{1.27}$$

and this is established in Chapter 5; much more is true. All we need here is that, on a suitable sequence of asymptotic density 1, the minimum ratio in (1.27) converges to 1, and consequently is attained for a large j. Of course the sequence on the right of (1.24) has increasing ratios.

Hence Theorems 12 and 13 should be regarded as useful and important tools in the fine study of divisors which are best possible as they stand. Nevertheless, for certain important problems they are inappropriate – and they cannot be 'improved'. We have to look at these things another way.

Notes on Chapter 1

§§1.1–1.2. An easy proof, based on the Turán–Kubilius inequality, of a slightly weaker result than Theorem 10 appears in Erdős (1979). It is remarkable that one does not have

$$\operatorname{card}\{k \leqslant \omega(n): \log_2 p_k(n) > k\} \sim \tfrac{1}{2}\log_2 n \quad \text{p.p.,} \tag{1}$$

although the weighted analogue holds:

$$\sum \{1/k: \log_2 p_k(n) > k\} \sim \tfrac{1}{2}\log_3 n \quad \text{p.p.} \tag{2}$$

In fact the left-hand side of (1) divided by $\log\log n$ has a distribution function equal to the arcsine law – see Erdős (1969).

For the sake of completeness, we have presented here a *direct* proof of the law of iterated logarithm stated in Theorem 12. That is we use only number theoretical estimates and we do not borrow any result from Probability Theory. One could also follow an *indirect* approach, as in Kubilius' book (Theorem 7.2), with the aim of applying Kolmogorov's classical law after having established a suitable strong approximation theorem. Since the necessary details are not quite provided there, we have decided to give a complete, self contained proof. For recent generalizations to other arithmetic functions, and to Strassen's form, see Manstavičius (1986a, b).

In his 1979 paper Erdős introduced the densities $\Lambda_k(d)$, resp. $\lambda_k(p)$, of those integers n such that $d_k(n) = d$, resp. $p_k(n) = p$. Although $p_k(n)$ is 'about' $\exp\exp k$ for most n, $\lambda_k(p)$ assumes its maximum value around $\exp(k/\log k)$, (there seems to be a slip about this in Erdős' paper, cf. Exercise 12). Similarly $\Lambda_k(d)$ is maximal for values of d much smaller than the 'normal' order $\exp(k^{1/\log 2})$: cf. Exercise 13.

According to Theorem 11, the expected value of $\omega(n, s, t) := \omega(n, t) - \omega(n, s)$ is $\log_2 t - \log_2 s$, but we need this to be large compared with $\sqrt{(\log_2 t \cdot \log_4 t)}$. Erdős (1969) proved that

$$\omega(n, s, t) \sim \log\left(\frac{\log t}{\log s}\right) \quad \text{p.p.}$$

uniformly for $s \leqslant t \leqslant n$ restricted so that the right-hand side exceeds $\psi(n)\log_3 n$, $\psi(n) \to \infty$; moreover this condition is necessary. Both papers contain other striking results on the normal distribution of prime factors. Bovey (1977) is also fundamental.

Exercises on Chapter 1

10. By setting up checkpoints at $t = t_0^{2^k}$, show that the sequence $\mathscr{B} = \mathscr{B}(\kappa, t_0)$ of integers n for which

$$\sup_{t \geqslant t_0} \left| \frac{\omega(n,t)}{\log_2 t} - 1 \right| > \kappa$$

has upper asymptotic density $\bar{\mathbf{d}}\mathscr{B} \ll_\kappa (\log t_0)^{-K}$, where $K = (1+\kappa)\log(1+\kappa) - \kappa$, uniformly for $t_0 \geqslant 3$.

11. Let $\lambda_k(p)$ denote the asymptotic density of the set of integers n such that $p_k(n) = p$. Show that

$$\lambda_k(p) = \frac{1}{p} \prod_{p' < p} \left(1 - \frac{1}{p'}\right) \sum_{\substack{\omega(m) = k-1 \\ P^+(m) < p}} m^{-1}$$

and deduce that, for $p \to \infty$, $\log_2 k = \mathrm{o}(\log_2 p)$,

$$\lambda_k(p) = \frac{\mathrm{e}^{-\gamma}}{p \log p} \frac{(\log_2 p + \mathrm{O}(\log_2 k))^{k-1}}{(k-1)!}$$

Deduce that, for given k, $\lambda_k(p)$ assumes its maximal value when $p = \exp\{(1 + \mathrm{o}(1))k/\log k\}$. How can this last statement be compatible with the law of the iterated logarithm for prime factors?

12. Let $\xi(t) \to \infty$ arbitrarily slowly and put $L(t) = \sqrt{(2\log_2 t \cdot \log_4 t)}$ $(t \geqslant t_0)$.

(a) Show that, p.p.x,

$$|\Omega(n,t) - \log_2 t| < (1+\varepsilon)L(t) \quad (\xi(x) < t \leqslant x).$$

(b) Put $D(n,t) = \mathrm{card}\{d \colon d|n, d \leqslant t\}$. Deduce from (a) that, p.p.x,

$$\log D(n,t)/\log 2 \leqslant \log_2 t + (1+\varepsilon)L(t) \quad (\xi(x) < t \leqslant x).$$

(c) Put $\bar{D}(n;t,u) = \mathrm{card}\{d \colon d|n, d > t, P^+(d) \leqslant u\}$. Show that, uniformly for $2 \leqslant t$, $u \leqslant x$ we have

$$\sum_{n \leqslant x} \bar{D}(n;t,u) \ll x \log u \cdot \exp\{-c_0 \log t/\log u\}.$$

(d) Put $\alpha = 1/\log 2$, $t_j = \exp j^\alpha$ $(j \geqslant 1)$ and $u_j = \exp\{c_0 j^\alpha/4\log j\}$ where c_0 is the constant appearing in part (c). Prove that, p.p.x,

$$\bar{D}(n;t_j,u_j) = 0 \quad (\xi(x) < j \leqslant (\log x)^{\log 2})$$

where $\xi(x) \to \infty$ arbitrarily slowly.

(e) Deduce from (d), using the trivial inequality $2^{\omega(n,u)} \leqslant D(n,t) + \bar{D}(n;t,u)$ $(n,t,u \geqslant 1)$, that, p.p.x,

$$\frac{\log D(n,t_j)}{\log 2} \geqslant \log_2 t_j - (1+\varepsilon)L(t_j) \quad (\xi(x) < j \leqslant (\log x)^{\log 2});$$

moreover, for at least one j in this range,

$$\frac{\log D(n, t_j)}{\log 2} \geqslant \log_2 t_j + (1-\varepsilon)L(t_j).$$

(f) Deduce Theorem 13 from part (e).

13. For $k, d \geqslant 1$ let $\Lambda_k(d)$ denote the density of those integers n for which $d_k(n) = d$. Set $j = 1 + [\log k/\log 2]$ and let N be the product of the first j primes. Find a lower bound for $\Lambda_k(d_k(N))$ and deduce that $\Lambda_k(d)$ is maximal when

$$k \leqslant d \leqslant \exp\left\{(1+o(1))\frac{\log k}{\log 2}\log_2 k\right\}.$$

Compare this result with Theorem 13.

2
Sieving by an interval

2.1 Introduction

Amongst the integer sequences defined by multiplicative constraints, certainly one of the simplest and most natural is the set of multiples of the integers lying in a given interval: this is a sieve problem, with the emphasis on the sifted rather than the unsifted numbers!

Let $H(x, y, z)$ denote the number of integers n not exceeding x which have at least one divisor d such that $y < d \leqslant z$. Estimates for $H(x, y, z)$ are needed in many arithmetical problems; in particular they provide one of the approaches to the propinquity of divisors.

Historically the problem about $H(x, y, z)$ was first introduced in a special but, as we shall see, important case in connection with the theory of *sets of multiples*. For any integer sequence $\mathscr{A}$, let

$$\mathscr{B}(\mathscr{A}) := \{ma : m \in \mathbb{Z}^+, a \in \mathscr{A}\}$$

be the set of multiples of $\mathscr{A}$. Such a set $\mathscr{B}$ may of course be generated by many sequences $\mathscr{A}$, but there is a unique, minimal sequence $\mathscr{A}'$ such that $\mathscr{B}(\mathscr{A}') = \mathscr{B}(\mathscr{A})$, characterized by the property that *no element of $\mathscr{A}'$ divides any other*. Sequences with this property are called primitive. In fact $\mathscr{A}'$ is the intersection of all the sequences that generate $\mathscr{B}$ (see the Notes, also Halberstam & Roth (1966), Chapter V). An example of a finite primitive sequence is

$$\mathscr{A}(y) := \{a : y < a \leqslant 2y\}$$

and the counting function of $\mathscr{B}(\mathscr{A}(y))$ is $H(x, y, 2y)$.

In the early thirties it was conjectured that any set of multiples had asymptotic density. A counterexample was constructed by Besicovitch (1934) (cf. Exercise 20), who proved as a key step

$$\liminf_{y \to \infty} \lim_{x \to \infty} x^{-1} H(x, y, 2y) = 0. \tag{2.1}$$

Erdős (1935) showed that liminf could be replaced by lim, and more than 40 years later gave another consequence of this strong form of (2.1), that the average order of Hooley's function

$$\Delta(n) := \max_{u} \operatorname{card} \{d : d \mid n, e^u < d \leqslant e^{u+1}\}$$

tends to infinity (cf. Hooley (1979), and Theorem 60).

One of the main results of the present chapter is a strong quantitative version of (2.1),

$$H(x, y, 2y) \leqslant \frac{x}{(\log y)^\delta \sqrt{\log_2 y}} \quad (3 \leqslant y \leqslant \sqrt{x}), \tag{2.2}$$

where the exponent

$$\delta = 1 - \frac{1 + \log_2 2}{\log 2} = 0.086\,07\ldots \tag{2.3}$$

will be shown to be optimal. In Exercise 22, the reader will see how to apply (2.2) in Erdős' method to obtain

$$\sum_{n \leqslant x} \Delta(n) \gg x \log_3 x. \tag{2.4}$$

Another topic for which $H(x, y, z)$ proves to be important is the study of relations between consecutive divisors – often the subject of Erdős' 'unconventional' problems. For instance, let

$$F(n; \theta) = \sum_{1 \leqslant i < \tau(n)} \theta(d_i/d_{i+1})$$

where $d_i = d_i(n)$ and $\theta \in C^2[0, 1]$ is given. Then (Erdős & Tenenbaum (1983)),

$$\sum_{n \leqslant x} F(n, \theta) = \theta(1) x \log x - 2 \int_1^{\sqrt{x}} \int_1^{\sqrt{x}} \varphi\left(\frac{y}{z}\right) H(x, y, z) \frac{\mathrm{d}y\,\mathrm{d}z}{z^2} + \mathrm{O}(x), \tag{2.5}$$

where $\varphi(t) = \theta'(t) + t\theta''(t)$. A detailed proof and several applications are to be found in Exercises 23–25.

Finally let D_n be the random variable whose distribution function is

$$F_n(v) := \frac{1}{\tau(n)} \operatorname{card}\{d\colon d|n, d \leqslant n^v\}. \tag{2.6}$$

In principle any question about the divisors of n can be posed in terms of F_n and, since we know from Theorem 14 that F_n does not converge weakly in any subsequence of $\mathbb{Z}^+$ of positive density, we must seek an alternative description of its limiting behaviour. A result of this sort is (Tenenbaum (1979)):

Theorem 20. *For each fixed pair of real numbers u, v* $(0 \leqslant u < v \leqslant 1)$, *the arithmetical function*

$$n \mapsto F_n(v) - F_n(u) = \frac{1}{\tau(n)} \operatorname{card}\{d\colon d|n, n^u < d \leqslant n^v\}$$

has a purely discrete distribution function all of whose step-points are dyadic rationals $a \cdot 2^{-b}$ $(a, b \in \mathbb{Z}^+)$.

In this application one needs upper bounds for $H(x, y, z)$ when y and z are fixed powers of x.

2.2 Statement of results concerning $H(x,y,z)$

There are essentially four possibilities for the asymptotic behaviour of this function depending on the values of y and z. We introduce some notation to help describe what happens.

We recall the constant δ of (2.3) and from (0.18) the function

$$Q(v) = v\log v - v + 1 \quad (v > 0).$$

We define

$$G(v) = \left\{ \begin{array}{ll} Q\left(\dfrac{1+v^+}{\log 2}\right) & (v \leqslant \log 4 - 1), \\ v & (v > \log 4 - 1), \end{array} \right\} \tag{2.7}$$

$$L(v) = \exp\{\sqrt{(\log v \cdot \log_2 3v)}\} \quad (v \geqslant 1).$$

Notice that Q is increasing when $v > 1$ and $Q(1/\log 2) = \delta$. Thus $G(v) = \delta$ for $v \leqslant 0$; for $v > 0$, G increases and since $G'(\log 4 - 1 - 0) = 1$ it is continuously differentiable. Moreover $G'(v) < 1$ for $v < \log 4 - 1$ so that in this range we have $G(v) > v$.

For $2 \leqslant y < z$ we define $u = u(y, z)$, $\beta = \beta(y, z)$, and $\xi = \xi(y, z)$ implicitly from the relations

$$z = y^{1+u} = y + y(\log y)^{-\beta}, \quad \beta = \log 4 - 1 + \frac{\xi}{\sqrt{\log_2 y}}. \tag{2.8}$$

Theorem 21. (i) *Let* $y \to \infty$, $z - y \to \infty$, $y \leqslant z \leqslant \sqrt{x}$, *and suppose in addition that* $\xi \to +\infty$. *Then*

$$H(x, y, z) = (1 + o(1))x\log^{-\beta} y. \tag{2.9}$$

(ii) *Let* $2 \leqslant y < z \leqslant \min(2y, \sqrt{x})$, *and* ξ *be bounded above. Then*

$$x(\log y)^{-G(\beta)}L(\log y)^{-c_1} \ll H(x, y, z) \ll x(\log y)^{-G(\beta)}/(1 + (-\xi)^+), \tag{2.10}$$

where c_1 *is an absolute constant.*

(iii) *There exists an absolute constant* c_2 *so that in the range* $y \geqslant 2$, $2y \leqslant z \leqslant \min(y^{3/2}, \sqrt{x})$ *we have*

$$xu^\delta L(1/u)^{-c_2} \ll H(x, y, z) \ll xu^\delta \frac{\log_2(3/u)}{\sqrt{\log(2/u)}}; \tag{2.11}$$

moreover, for each fixed $B \geqslant 2$, *if* $z \leqslant By$ *we may strike out the factor* $\log_2(3/u)$ *provided we replace* $\ll$ *by* $\ll_B$ *on the right.*

(iv) *Uniformly for* $2 \leqslant y \leqslant z \leqslant x$ *we have*

$$H(x, y, z) = x\left(1 + O\left(\frac{\log y}{\log z}\right)\right). \tag{2.12}$$

Part (iv) is very simple, and it is non-trivial only if $z > y^c$ for a suitably large c. However, $H(x, y, z)$ is an increasing function of z and so for $y^{3/2} < z \leqslant y^c$ we certainly have $H(x, y, z) \asymp x$, (2.11) providing the lower bound.

The following heuristic explanation of the behaviour of $H(x, y, z)$ may be helpful.

When z is very close to y, the different conditions $n \equiv 0 \pmod d$, $y < d \leqslant z$, are sufficiently independent (the various of g.c.d.'s being small) for us to have (we assume $y \to \infty$, β large)

$$H(x, y, z) \sim \sum_{y<d\leqslant z} \left[\frac{x}{d}\right] \sim x \log(z/y) \sim x(\log y)^{-\beta}.$$

This is the state of affairs in (2.9). In fact the threshold of dependence occurs at $\beta = \log 4 - 1$: when β is smaller than this critical value the interaction between the implied congruence conditions has a perceptible effect on the exponent of $\log y$ which is now smaller than $-\beta$. The factor lost due to dependence increases as z increases towards $2y$, when it is $(\log y)^{\delta + o(1)}$.

When $z > 2y$ a new phenomenon appears owing to the influence of the 'small' primes $\leqslant z/y$. By Theorem 07, we may asssume that usually

$$n(z/y) := \prod_{\substack{p^\alpha \| n \\ p \leqslant z/y}} p^\alpha$$

is itself fairly small: the small prime factors of n may be expected to affect not so much the existence of divisors in $(y, z]$ but their number, given that this is positive, and this is irrelevant. So we are concerned largely with divisors made up of primes from $(z/y, z] = (y^u, y^{1+u}]$, and $1/\log y$ gives way to u when we pass from (2.10) to (2.11). The fact that δ now remains constant reveals some sort of saturation in the dependence.

Finally, when z becomes so large that u itself is large, almost all integers have not just a divisor but a prime factor in $(y, z]$. This simple remark is half-way to a proof of the easy, but as we shall see practically optimal formula (2.12).

It is worthwhile to grasp at the outset why the threshold of dependence is at $\beta = \log 4 - 1$. It is well-known that the sum $\sum\{\tau(n): n \leqslant x\}$ is dominated by the n with $\Omega(n) \sim 2\log_2 x$. Similarly (and both these statements may readily be justified by the method of §0.5),

$$\sum_{n \leqslant x} \tau(n; y, z) \tag{2.13}$$

is dominated by those n for which $\Omega(n, z) \sim 2 \log\log z$.

Now (2.13) is a trivial upper bound for $H(x, y, z)$. But in the case of quasi-independence we must have

$$x^{-1}H(x, y, z) \sim 1 - \prod_{y<d\leqslant z}\left(1 - \frac{1}{d}\right) = 1 - \frac{[y]}{[z]} \sim x^{-1}\sum_{n\leqslant x}\tau(n; y, z) \tag{2.14}$$

(we are assuming $z - y = o(y)$, $z \leqslant \sqrt{x}$), and so the dominant contribution in (2.13) must come from terms with $\tau(n; y, z) = 0$ or 1. If, as is acceptable for the present discussion, we identify $\tau(n; 0, z)$ with $2^{\Omega(n,z)}$ and further assume that the set $\{\log d: d|n, d \leqslant z\}$ is more or less evenly distributed over $[0, \log z]$, we see that $\tau(n; y, z)$ should be about

$$2^{\Omega(n,z)}\frac{\log(z/y)}{\log z} = (\log y)^{\log 4 - 1 - \beta + o(1)}$$

for the dominant integers n in (2.13). The reader should see Erdős (1967), where $\log 4 - 1$ is also a threshold.

Let us write

$$H(x, y, z) =: \eta(x; y, z)\sum_{n\leqslant x}\tau(n; y, z).$$

Then $\eta(x; y, z) \leqslant 1$, moreover if we combine the upper bound (2.10) and (2.42) below, we see that for $\xi = O((\log_2 y)^{1/6})$ we have

$$\eta(x; y, z) \ll (1 + (-\xi)^+)^{-1}\exp(-((-\xi)^+)^2/\log^2 4).$$

Equally, the proof of (2.9) would give a bound for $1 - \eta(x; y, z)$ for large ξ.

Conjecture. *There exists a probability distribution function F such that, if $x, y, z \to \infty$ with $\xi \in \mathbb{R}$ constant,*

$$\eta(x, y, z) \to F(\xi).$$

2.3 More applications

First we return to the sequence of random variables D_n. We have already mentioned that the upper bound in (2.11) is vital for the proof of Theorem 20: an immediate corollary of that result and (2.11) itself is

Theorem 22. *Let t, u be real numbers, $t \geqslant 1$, $0 \leqslant u \leqslant 1$. Then the sequence of the integers n which have at least one divisor d such that*

$$n^{(1-u)/t} < d \leqslant n^{1/t} \tag{2.15}$$

possesses asymptotic density $h(u, t)$; furthermore for $t \geqslant 2$, $0 \leqslant u \leqslant 1$ we have

$$u^\delta L(1/u)^{-c_4} \ll h(u, t) \ll u^\delta\frac{\log_2(3/u)}{\sqrt{\log(2/u)}}. \tag{2.16}$$

(The restriction $t \geqslant 2$ is unimportant because of the symmetry of the divisors about $\sqrt{n}$.)

If we replace 'divisor' by 'prime factor' above, it is much easier to establish the existence of the density – such a result has probably been known implicitly for a long time. (For a complete proof see Tenenbaum (1980).) A famous special case concerns the integers n such that $P^+(n) \leqslant n^{1/v}$: the density is $\rho(v)$ where ρ is Dickman's function defined by

$$\left.\begin{aligned} \rho(v) &= 1 \quad (0 < v \leqslant 1), \\ v\rho'(v) &= -\rho(v-1) \quad (v > 1). \end{aligned}\right\} \tag{2.17}$$

Prime factors are easier to deal with because Buchstab's identity is available: this reads

$$\Psi(x, y) = 1 + \sum_{p \leqslant y} \Psi\left(\frac{x}{p}, p\right) \tag{2.18}$$

in the Dickman problem. There is nothing analogous to (2.18) for divisors. A typical feature of the proofs of Theorems 20 and 22 is that a relatively sharp upper bound for the counting function is needed prior to the proof that the density exists. We prove Theorem 22 in full detail in the next section, assuming (2.11).

Our second application also concerns the random variables D_n. It consists of the rather surprising 'incertitude principle' already stated in Theorem 14: *if $\mathscr{A}$ is an integer sequence such that $\{D_n: n \in \mathscr{A}\}$ has a limit law, then* $\mathbf{d}\mathscr{A} = 0$. We are now in a position to prove this as an easy consequence of Theorem 22.

Proof of Theorem 14. This is by contradiction: we assume $\bar{\mathbf{d}}\mathscr{A} > 0$. We may assume that $\tau(n) \to \infty$ $(n \in \mathscr{A})$ as the complementary set has zero density. Now let F be the distribution function of the limit law so that by hypothesis

$$\lim_{\substack{n \to \infty \\ n \in \mathscr{A}}} F_n(u) = F(u) \quad (u \in \mathscr{C}(F)). \tag{2.19}$$

Recall that F is *symmetric* if $F(u) + F(1-u) = 1$ $(u \in \mathscr{C}(F))$: this is the case because each F_n in (2.19) is symmetric. In particular if u is a *point of increase* of F (i.e. $F(u+\varepsilon) - F(u-\varepsilon) > 0$ for all $\varepsilon > 0$), so is $1-u$. We are going to show that F has at least one point of increase and hence at least one in $[0, \frac{1}{2}]$. This follows from the Radon–Nykodym decomposition $\mathrm{d}F = \nu + f\,\mathrm{d}u$, where ν is a singular measure and $f \geqslant 0$. Indeed if $\nu \not\equiv 0$ the quantity

$$\liminf_{\varepsilon \to 0} \frac{1}{2\varepsilon}(F(u+\varepsilon) - F(u-\varepsilon)) \tag{2.20}$$

is infinite for ν-almost all, and hence for at least one, u. Alternatively, if $\nu \equiv 0$, (2.20) is identically $f(u)$ and is positive for some u. Thus F does have a point of increase in $[0, \frac{1}{2}]$.

We consider two cases according as u may be taken in $(0, \frac{1}{2}]$ or not. In case (i), for arbitrarily small ε both $u \pm \varepsilon \in \mathscr{C}(F)$ and

$$\operatorname{card}\{d: d|n, u^{u-\varepsilon} < d \leqslant n^{u+\varepsilon}\} \sim \tau(n)(F(u+\varepsilon) - F(u-\varepsilon))$$

as $n \to \infty$ through $\mathscr{A}$. As $\tau(n) \to \infty$, the left-hand side above is $\geqslant 1$ for large $n \in \mathscr{A}$. By Theorem 22, we deduce that

$$\bar{\mathbf{d}}\mathscr{A} \leqslant h\left(\frac{2\varepsilon}{u+\varepsilon}, \frac{1}{u+\varepsilon}\right) = \mathrm{O}(\varepsilon^{\delta}) \quad (\varepsilon \to 0),$$

and $\bar{\mathbf{d}}\mathscr{A} = 0$ as required. In case (ii), $\mathrm{d}F$ concentrates all its mass at the end-points $u = 0, 1$: by the symmetry, we must have

$$F(u) = \begin{cases} 0 & (u < 0), \\ \frac{1}{2} & (0 \leqslant u < 1), \\ 1 & (u \geqslant 1), \end{cases}$$

and so, for fixed $u > 0$, $F_n(u) \geqslant \frac{1}{3}$ for all sufficiently large $n \in \mathscr{A}$, and

$$\bar{\mathbf{d}}\mathscr{A} \leqslant \limsup_{x \to \infty} 3x^{-1} \sum_{n \leqslant x} F_n(u). \tag{2.21}$$

The result of Deshouillers, Dress & Tenenbaum (1979) (1.25) implies that the right-hand side of (2.21) is equal to $(6/\pi)\arcsin\sqrt{u}$, and since u may be taken arbitrarily small we again have $\bar{\mathbf{d}}\mathscr{A} = 0$. (We leave to the reader, as an exercise on Theorem 01, a self-contained proof that the right-hand side of (2.21) is $\ll \sqrt{u}$.) This completes the proof.

Our third application concerns a question of Linnik and Vinogradov: to determine as exactly as possible the number $A(x)$ of *distinct* products $ab \leqslant x$ such that $a \leqslant \sqrt{x}$, $b \leqslant \sqrt{x}$. We obtain the following result, sharpening that of Erdős (1960).

Theorem 23. *For $x \geqslant 3$, we have*

$$\frac{x}{(\log x)^{\delta}} L(\log x)^{-c_5} \ll A(x) \ll \frac{x}{(\log x)^{\delta}} (\log_2 x)^{-1/2}$$

with δ as in (2.3).

Proof. The result follows from Theorem 21 (iii) and

$$H\left(\frac{x}{4}, \frac{\sqrt{x}}{4}, \frac{\sqrt{x}}{2}\right) \leqslant A(x) \leqslant \sum_{k=0}^{\infty} H\left(\frac{x}{4^k}, \frac{\sqrt{x}}{4^{k+1}}, \frac{\sqrt{x}}{2^k}\right) \quad (x \geqslant 4). \tag{2.22}$$

To prove (2.22), we establish inclusions between the various sets. If n is counted on the left, then $n = md \leqslant x/4$ with $\sqrt{x}/4 < d \leqslant \sqrt{x}/2$ say, whence $m \leqslant \sqrt{x}$ and n is counted by $A(x)$. If n is counted by $A(x)$ let k be such that $x/4^{k+1} < n \leqslant x/4^k$. By hypothesis $n = md$, $\max(m, d) \leqslant \sqrt{x}$. Suppose for instance that $d \leqslant m$. Then $d \leqslant \sqrt{n} \leqslant \sqrt{x}/2^k$, $d = n/m \geqslant n/\sqrt{x} \geqslant \sqrt{x}/4^{k+1}$ and n is counted on the right.

In the fourth application we consider the sums

$$M(n, y) := \sum_{\substack{d \mid n \\ d \leqslant y}} \mu(d)$$

For $n > 1$, $M(n, y) = 0$ when $y \geqslant n$, but one may be interested in $M(n, y)$ as a function of $y \leqslant n$. For example it occurs in Vaughan's method for binary forms (Vaughan (1980)). Maier (1987) has established that

$$M(n) := \max_{1 \leqslant y \leqslant n} |M(n, y)| \to \infty \quad \text{p.p.}$$

and we have the following result sharpening that of Erdős & Hall (1980).

Theorem 24. *For $y > 1$, let ε_y denote the asymptotic density of the sequence $\{n: M(n, y) \neq 0\}$. Then $\varepsilon_y \ll (\log y)^{-\delta}$.*

For a (sketch) proof see Exercise 211. We obtain the slightly inferior exponent $-\delta/(1+\delta)$ here.

The density exists, the condition $M(n, y) \neq 0$ involving only congruences (mod d), $d \leqslant y$. Let $m = \prod\{p: p \mid n\}$ be the core of n and $q = P^-(n) = P^-(m)$. Then

$$M(n, y) = \sum_{\substack{d \mid m/q \\ d \leqslant y}} \mu(d) + \sum_{\substack{d \mid m/q \\ qd \leqslant y}} \mu(qd) = \sum_{\substack{d \mid m/q \\ y/q < d \leqslant y}} \mu(d),$$

and so $M(n, y) \neq 0$ implies n has a divisor $d \in (y/P^-(n), y]$. Let $\alpha \in (0, 1)$ be at our disposal. Either $P^-(n) > \exp\{(\log y)^\alpha\}$ or n is counted by $H(x, y\exp\{-(\log y)^\alpha\}, y)$. The number of such n is

$$\ll x\{(\log y)^{-\alpha} + (\log y)^{-\delta(1-\alpha)}\}$$

by Theorem 06 and (2.11) respectively. We choose $\alpha = \delta/(1+\delta)$.

2.4 Proof of Theorem 22

We employ Theorem 21 to establish the existence of the density $h(u, t)$ of integers n with at least one divisor satisfying (2.15), for all $t \geqslant 1$, $u \in [0, 1]$. We notice that it is sufficient to deal with the range $t \geqslant 2$, $u \in (0, 1)$. The cases $u = 0$ or 1 are trivial since $h(0, t) = 0$, $h(1, t) = 1$ for all $t \geqslant 1$, and in view of the symmetry of the divisors about $\sqrt{n}$ we also have

$$h(u,t)=\begin{cases} h\left(\dfrac{u}{t+u-1},\dfrac{t}{t+u-1}\right) & (1\leqslant t\leqslant 2-2u),\\ h\left(\dfrac{2-t}{t},2\right) & (2-2u\leqslant t\leqslant 2-u),\\ h\left(\dfrac{t+2u-2}{t},2\right) & (2-u\leqslant t<2).\end{cases} \tag{2.23}$$

We assume $t\geqslant 2$, $u\in(0,1)$ and put

$$H(x):=\operatorname{card}\{n\leqslant x:\exists d\colon d|n,\, n^{(1-u)/t}<d\leqslant n^{1/t}\}.$$

It is natural to expect that

$$H(x)=H(x,y,z)+\mathrm{o}(x)\quad (x\to\infty) \tag{2.24}$$

with

$$y=x^{(1-u)/t},\quad z=x^{1/t}, \tag{2.25}$$

and (2.24) is our first goal. Once its existence is established, the bounds (2.16) for $h(u,t)$ will follow from (2.11) and (2.24).

Let $n\in(x/\log x, x]$ be counted by one, but not both, of $H(x)$ and $H(x,y,z)$. Then n has a divisor in $((x/\log x)^{\alpha}, x^{\alpha}]$ with $\alpha=(1-u)/t$ or $1/t$, and by (2.11) the number of such n is $\ll_{u,t} x(\log x)^{-\delta}$. Hence (2.24) (with this upper bound for the remainder). It remains to establish the existence of

$$\lim_{x\to\infty} x^{-1}H(x,y,z)=h(u,t) \tag{2.26}$$

where here and throughout this section y and z are as in (2.25).

Eratosthenes' sieve gives

$$H(x,y,z)=\sum_{k=1}^{\infty}(-1)^{k-1}\sum_{y<d_1<\cdots<d_k\leqslant z}\left[\frac{x}{[d_1,d_2,\ldots,d_k]}\right]$$

but it would be very difficult to use this for a proof – for fixed k the inner sum is not $\mathrm{O}(x)$. We should have more hope if the d_i were restricted to be primes or had a bounded number of prime factors and with this in mind we define an auxiliary quantity $H_\varepsilon(x,y,z)$ which counts the integers $n\leqslant x$ with a divisor $d\in(y,z]$ satisfying $P^-(d)>y^\varepsilon$. We show that

$$\lim_{\varepsilon\to 0}\limsup_{x\to 0} x^{-1}(H(x,y,z)-H_\varepsilon(x,y,z))=0, \tag{2.27}$$

again by means of (2.11). Indeed, let n be counted by $H(x,y,z)$ but not $H_\varepsilon(x,y,z)$, and let d be the smallest divisor of n in $(y,z]$. Then necessarily $d/P^-(d)\leqslant y$, $P^-(d)\leqslant y^\varepsilon$, that is $d\in(y,y^{1+\varepsilon}]$, and n is counted by $H(x,y,y^{1+\varepsilon})$. Thus

$$H(x,y,z)-H_\varepsilon(x,y,z)\leqslant H(x,y,y^{1+\varepsilon})\ll \varepsilon^\delta x$$

by (2.11), and (2.27) follows. It will therefore be sufficient to show that

$$\lim_{x\to\infty} x^{-1}H_\varepsilon(x,y,z)=:h_\varepsilon(u,t) \tag{2.28}$$

exists for each fixed positive ε.

By the sieve of Eratosthenes,

$$H_\varepsilon(x,y,z)=\sum_{k=1}^{\infty}(-1)^{k-1}B_k(x)$$

where

$$B_k(x)=\sum_{\substack{y<d_1<\cdots<d_k\leqslant z\\ P^-(d_i)>y^\varepsilon}}\left[\frac{x}{[d_1,\ldots,d_k]}\right].$$

If $\{d_1,d_2,\ldots,d_k\}$ is a k-tuple of integers with non-zero contribution to $B_k(x)$, we must have

$$d:=[d_1,\ldots,d_k]\leqslant x,\quad P^-(d)>y^\varepsilon$$

whence

$$k\leqslant\tau(d)\leqslant 2^{\Omega(d)}<2^{t/\varepsilon(1-u)}. \tag{2.29}$$

Hence the limit $h_\varepsilon(u,t)$ in (2.28) exists if, for each fixed k, $\lim B_k(x)/x$ exists.

Let $\rho(d)$ denote the number of representations of d in the form $d=[d_1,d_2,\ldots,d_k]$ with $y<d_1<\cdots<d_k\leqslant z$. We have

$$B_k(x)=\sum_{\substack{d=1\\ P^-(d)>y^\varepsilon}}^{\infty}\rho(d)\left[\frac{x}{d}\right]=x\sum_{\substack{d\leqslant x\\ P^-(d)>y^\varepsilon}}\frac{\rho(d)}{d}+O\Bigg(\sum_{\substack{d\leqslant x\\ P^-(d)>y^\varepsilon}}\rho(d)\Bigg),$$

and the error term is $\ll x/\log x$ because $\Omega(d)\ll_{t,u,\varepsilon}1$, $\rho(d)\leqslant\binom{\tau(d)}{k}\ll_{t,u,\varepsilon}1$.

The main term may be re-written as

$$x\sum_{r=1}^{\infty}\sum_{j=1}^{\infty}r{\sum}'\frac{1}{p_1p_2\cdots p_j} \tag{2.30}$$

where the dash means that summation is over all j-tuples of primes $\{p_1,p_2,\ldots,p_j\}$ such that

$$\left.\begin{aligned} &y^\varepsilon<p_1\leqslant p_2\leqslant\cdots\leqslant p_j\leqslant x,\\ &p_1p_2\cdots p_j\leqslant x,\\ &\rho(p_1\cdots p_j)=r.\end{aligned}\right\} \tag{2.31}$$

Also r and j are bounded, independently of x, and for every pair $\{r,j\}$ we may re-write (2.31) in the form

$$\chi\left(\frac{\log p_1}{\log x},\frac{\log p_2}{\log x},\ldots,\frac{\log p_j}{\log x}\right)=1$$

where χ is the characteristic function of a subset of $[\varepsilon(1-u)/t,1]^j$ defined by a bounded number of linear inequalities. The following lemma therefore implies that, as $x \to \infty$, $\sum'(1/p_1p_2\ldots p_j)$ tends to a limit, as does $x^{-1}B_k(x)$ via (2.30).

Lemma 22.1. *Let $\chi:(\mathbb{R}^+)^j \to \mathbb{R}$ be Riemann integrable with support contained in a compact subset of $(\mathbb{R}^+)^j$. Then we have*

$$\lim_{x\to\infty} \sum_{p_1,p_2,\ldots,p_j} \frac{\chi\left(\frac{\log p_1}{\log x},\ldots,\frac{\log p_j}{\log x}\right)}{p_1p_2\ldots p_j} = \int_0^\infty \cdots \int_0^\infty \chi(u_1,\ldots,u_j)\frac{du_1}{u_1}\cdots\frac{du_j}{u_j}.$$

Proof. On the hypercube $[a,b]^j$ $(0<a<b)$, define two measures ν and ν_x, the second depending on the parameter $x\to\infty$ by

$$d\nu(u_1,\ldots,u_j) = \frac{du_1}{u_1}\frac{du_2}{u_2}\cdots\frac{du_j}{u_j}$$

$$d\nu_x(u_1,\ldots,u_j) = \prod_{i\leqslant j} x^{-u_i}\, d\pi(x^{u_i}).$$

By the Prime Number Theorem and partial integration we readily obtain that for any product domain $P=\prod[a_i,b_i]\subseteq[a,b]^j$ we have

$$\lim_{x\to\infty}\int_P d\nu_x(u_1,\ldots,u_j) = \int_P d\nu(u_1,\ldots,u_j).$$

Hence the family of positive measures ν_x $(x>0)$ is uniformly bounded (take $P=[a,b]^j$) and converges weakly to ν on the set of finite linear combinations of characteristic functions of product domains. Therefore it also converges weakly to ν on the set of continuous functions supported on $[a,b]^j$. Since χ is Riemann integrable there exist such functions f and g with

$$f\leqslant\chi\leqslant g,\quad \int_{[a,b]^j}(g-f)\,d\nu\leqslant\varepsilon$$

for arbitrary $\varepsilon>0$. We obtain from

$$\int f\,d\nu + o(1) = \int f\,d\nu_x \leqslant \int \chi\,d\nu_x \leqslant \int g\,d\nu_x = \int g\,d\nu + o(1)$$

and

$$\int f\,d\nu \leqslant \int\chi\,d\nu \leqslant \int g\,d\nu$$

that

$$\left|\int\chi\,d\nu_x - \int\chi\,d\nu\right| \leqslant \varepsilon + o(1)$$

and we let $x\to\infty$, $\varepsilon\to 0$. The result follows.

2.5 Proof of Theorem 21 (i) – small z

In this section, we put $\eta=(z-y)/y=(\log y)^{-\beta}$ and our aim is to prove

$$H(x,y,z)=(1+o(1))\eta x \tag{2.32}$$

under the conditions $y\to\infty$, $z-y\to\infty$, $y\leqslant z\leqslant\sqrt{x}$, and

$$(\beta+1-\log 4)\sqrt{\log_2 y}\to\infty.$$

Under these hypotheses

$$\sum_{n\leqslant x}\tau(n;y,z)=\sum_{y<d\leqslant z}\left[\frac{x}{d}\right]\sim\eta x \tag{2.33}$$

from which the upper bound in (2.32) is clear.

For the lower bound we consider two cases. First, suppose that $\eta\log^2 y<1$. We have

$$\sum_{n\leqslant x}\tau(n;y,z)^2\leqslant\sum_{y<d\leqslant z}\frac{x}{d}+2x\sum_{y<d<d'\leqslant z}\frac{1}{[d,d']}. \tag{2.34}$$

The first sum on the right is $\sim x\log(z/y)\sim\eta x$. In the second we put $m=(d,d')$, $d=mt$, $d'=mt'$ and observe that $y<d<d'\leqslant z$ implies that $t'/(t'-1)\leqslant t'/t<1+\eta$ whence $t'>1+1/\eta$ and $m\leqslant z/t'\leqslant\eta y$. Therefore

$$\begin{aligned}\sum_{y<d<d'\leqslant z}\frac{1}{[d,d']}&\leqslant\sum_{m\leqslant\eta y}\frac{1}{m}\sum_{y/m<t<t'\leqslant z/m}\frac{1}{tt'}\\&\leqslant\sum_{m\leqslant\eta y}\frac{1}{m}\left(\log\frac{z}{y}+O\left(\frac{m}{y}\right)\right)^2\ll\eta^2\log y.\end{aligned}$$

We insert these estimates into (2.34) to obtain

$$\sum_{n\leqslant x}\tau(n;y,z)^2\leqslant(1+o(1))\eta x+O(\eta^2 x\log y)\sim\eta x$$

(because $\eta\log y<1/\log y=o(1)$), and so Cauchy's inequality

$$\left(\sum_{n\leqslant x}\tau(n;y,z)\right)^2\leqslant H(x,y,z)\sum_{n\leqslant x}\tau(n;y,z)^2 \tag{2.35}$$

yields the required lower bound for (2.32).

Now consider the second case when $\eta\log^2 y\geqslant 1$. Let χ denote the characteristic function of the set of integers satisfying

$$\Omega(n,y)\leqslant Y(y):=2\log_2 y+\psi(y)\sqrt{\log_2 y}, \tag{2.36}$$

where $\psi(y)\to\infty$ as $y\to\infty$, $\psi(y)\ll(\log_2 y)^{1/6}$. We have, uniformly for $1\leqslant v\leqslant v_0<2$, that

$$\begin{aligned}\sum_{n\leqslant x}(1-\chi(n))\tau(n;y,z)&\leqslant\sum_{n\leqslant x}v^{\Omega(n,y)-Y(y)}\tau(n;y,z)\\&\ll xv^{-Y(y)}(\log y)^{v-1}\sum_{y<d\leqslant z}\frac{v^{\Omega(d)}}{d}.\end{aligned}$$

Theorem 04 gives, for every $N\in\mathbb{Z}^+$,

$$\sum_{y<d\leqslant z}\frac{v^{\Omega(d)}}{d}\ll_N(\log y)^{v-1}\log\frac{z}{y}+(\log y)^{v-N-1} \tag{2.37}$$

and we put $N=2$, so that the right-hand side is $\ll\eta(\log y)^{v-1}$ because $\eta\log^2 y\geqslant 1$. So we have

$$\sum_{n\leqslant x}(1-\chi(n))\tau(n;y,z)\ll\eta x v^{-Y(y)}(\log y)^{2v-2}$$

and we put $v=Y(y)/2\log_2 y$: the right-hand side is

$$\ll\eta x\exp(-\tfrac{1}{4}\psi(y)^2+O(\psi(y)^3/\sqrt{\log_2 y}))=o(\eta x).$$

Next,

$$H(x,y,z)\geqslant\sum_{\substack{n\leqslant x\\ \tau(n;y,z)=1}}1\geqslant\sum_{n\leqslant x}\chi(n)\left\{\tau(n;y,z)-2\binom{\tau(n;y,z)}{2}\right\}$$
$$\geqslant(1+o(1))\eta x-2S$$

say, where

$$S:=\sum_{n\leqslant x}\chi(n)\binom{\tau(n;y,z)}{2}\leqslant\sum_{n\leqslant x}v^{\Omega(n,y)-Y(y)}\sum_{\substack{d,d'|n\\ y<d<d'\leqslant z}}1,$$

provided $v\in(0,1]$. In the inner sum we introduce the g.c.d. $m=(d,d')$ and remark that, as before, for a non-empty sum we must have $m\leqslant\eta y$. We invert summations and apply Theorem 01 to obtain

$$S\ll xv^{-Y(y)}\sum_{m\leqslant\eta y}\frac{v^{\Omega(m)}}{m}(\log 2m)^{v-1}\left(\sum_{y/m<t\leqslant z/m}\frac{v^{\Omega(t,y)}}{t}\right)^2. \tag{2.38}$$

We claim that, uniformly for $0<v\leqslant 1$ and $m\leqslant\eta y$, we have

$$\sum_{y/m<t\leqslant z/m}\frac{v^{\Omega(t)}}{t}\ll\eta\left(\log\frac{1}{\eta}\right)^{1-v}\left(\log\frac{y}{m}\right)^{v-1}. \tag{2.39}$$

There are two cases according as $y/m\leqslant\eta^{-3/2}$ or not. In the first case, the trivial bound $\sum t^{-1}\ll\log(z/y)+m/y\ll\eta$ suffices. In the second case, we note that $(z-y)/m>(y/m)^{1/3}$ and we apply Shiu's result, Theorem 03.

For $t\leqslant y^2$, $\Omega(t,y)\geqslant\Omega(t)-1$. We insert (2.39) into (2.38) to deduce that

$$S\ll_v xv^{-Y(y)}\eta^2\left(\log\frac{1}{\eta}\right)^{2-2v}\sum_{m\leqslant\eta y}\frac{v^{\Omega(m)}}{m}(\log 2m)^{v-1}\left(\log\frac{y}{m}\right)^{2v-2}$$

and we put $v=\frac{1}{2}$. By partial summation, the inner sum is $\ll\log_2 y/\log y$, whence

$$S\ll x\eta^2(\log y)^{\log 4-1}2^{\psi(y)\sqrt{\log_2 y}}(\log_2 y)^2. \tag{2.40}$$

We put $\psi(y)=\min(\xi,(\log_2 y)^{1/6})$, and we have $S=o(\eta x)$. This gives the desired result.

2.6 Proof of Theorem 21 (ii), (iii) – upper bounds

Lemma 21.1. *Let $\Pi'_k(y)$ denote the number of integers $\leqslant y$ with exactly k odd prime factors (counted according to multiplicity). Then for any fixed c, $z > y + y(\log y)^{-c}$, and uniformly for $0 \leqslant k \leqslant (3-\varepsilon)\log_2 y$, we have*

$$\Pi'_k(z) - \Pi'_k(y) \ll_{\varepsilon,c} \frac{z-y}{\log y}\cdot\frac{(\log_2 y)^k}{k!}$$

Proof. We need a modification of Theorem 04 in which Ω' counts odd prime factors: for $|v| \leqslant 3-\varepsilon$, $x \geqslant 2$, $N \in \mathbb{Z}^+$, we have

$$\sum_{n\leqslant x} v^{\Omega'(n)} = \int_2^x \sum_{\nu=0}^{N} a'_\nu(v)(\log t)^{v-1-\nu}\,\mathrm{d}t + \mathrm{O}_{\varepsilon,N}(x(\log x)^{\mathrm{Re}\,v-2-N}) \qquad (2.41)$$

where $a'_\nu(v) \ll_{\varepsilon,N} 1$. Now put $v = r\mathrm{e}^{i\theta}$, $r \leqslant 3-\varepsilon$. We have

$$\begin{aligned}\{\Pi'_k(z) - \Pi'_k(y)\}r^k &= \frac{1}{2\pi}\int_0^{2\pi} \mathrm{e}^{-ik\theta} \sum_{y<n\leqslant z} v^{\Omega'(n)}\,\mathrm{d}\theta\\ &\ll_{\varepsilon,N} \int_y^z\int_0^{2\pi} (\log t)^{r\cos\theta - 1}\,\mathrm{d}\theta + z(\log y)^{r-2-N}\\ &\ll_{\varepsilon,N} (z-y)\frac{(\log y)^{r-1}}{\sqrt{(r\log_2 y)}} + z(\log y)^{r-2-N}\end{aligned}$$

and we put $r = k/(\log_2 y)$ and employ Stirling's formula. We take $N = [c]$.

Now we consider the upper bound in (2.10). If $\xi \geqslant -1$ we observe simply that

$$H(x,y,z) \leqslant \sum_{n\leqslant x} \tau(n;y,z) \ll x(\log y)^{-\beta}.$$

Now if $\beta < \log 4 - 1$, hence $-1 \leqslant \xi < 0$, we have

$$G(\beta) = \beta + \frac{(\log 4 - 1 - \beta)^2}{\log^2 4} + \mathrm{O}((\log 4 - 1 - \beta)^3) \qquad (2.42)$$

by Taylor's theorem: notice that $Q(2) = \log 4 - 1$, $Q'(2) = \log 2$. Thus in fact

$$H(x,y,z) \ll x(\log y)^{-G(\beta)}$$

which is the result stated. Next assume $\xi < -1$. We require Theorem 08, and we set $E = \{p: 3 \leqslant p < z\}$. Recall that $E(x) := \sum\{p^{-1}: p\in E,\ p\leqslant x\}$. (In this application the condition $p \leqslant x$ is of course redundant.) We split the integers counted by $H(x,y,z)$ into two classes with cardinalities H_1 and H_2 say, according as $\Omega(n,E) > \lambda E(x)$ or not. Here $\lambda = (1+\beta)/\log 2$. We have

$$\begin{aligned}H_1 &\leqslant \sum_{\lambda E(x) < k \leqslant 2.5\log_2 z} \operatorname{card}\{n: n\leqslant x, \Omega(n,E) = k\} + \sum_{\substack{n\leqslant x\\ \Omega(n,E) > 2.5\log_2 z}} 1\\ &\ll x\mathrm{e}^{-E(x)} \sum_{k>\lambda E(x)} E(x)^k/k! + x(\log z)^{-Q(2.5)}\end{aligned}$$

by Theorem 08 and an obvious variant of Exercise 04 respectively. Since $E(x) = \log_2 y + O(1)$, Theorem 09 yields

$$H_1 \ll x(\log y)^{-Q(\lambda)}(\log_2 y)^{-1/2}$$

which is within the upper bound stated. Next

$$H_2 \leqslant \sum_{\substack{n \leqslant x \\ \Omega(n,E) \leqslant \lambda E(x)}} \tau(n; y, z) \leqslant \sum_{h+k \leqslant \lambda E(x)} \sum_{\substack{y<d\leqslant z \\ \Omega(d,E)=h}} \sum_{\substack{m \leqslant x/d \\ \Omega(m,E)=k}} 1.$$

For $d \leqslant z \leqslant \sqrt{x}$ we note that $E(x/d) = E(x)$, and $\Omega(d, E) = \Omega'(d)$. Theorem 08 and Lemma 21.1 give

$$\begin{aligned} H_2 &\ll x e^{-E(x)} \sum_{h+k \leqslant \lambda E(x)} \frac{1}{y}(\Pi'_h(z) - \Pi'_h(y)) E(x)^k/k! \\ &\ll x(\log y)^{-1-\beta} e^{-E(x)} \sum_{l \leqslant \lambda E(x)} 2^l E(x)^l/l! \\ &\ll (1-\lambda/2)^{-1} x (\log y)^{-1-\beta} e^{E(x) - 2Q(\lambda/2)E(x)} E(x)^{-1/2} \end{aligned}$$

by Theorem 09. Since $2Q(\lambda/2) + \beta = Q(\lambda) = G(\beta)$, H_2 is also within the bound stated in (2.10). This establishes this part of the theorem.

The upper bound in part (iii), in which $2y < z \leqslant y^{3/2}$, is treated similarly, but the technical details are slightly more complicated because of the effect of the primes $\leqslant z/y$. We give a brief indication of the necessary modifications here, and refer the reader to Tenenbaum (1984) for a complete proof.

We now have $E = \{p: z/y < p \leqslant z\}$, and as before there are two classes of integers, split according as $\Omega(n, E) > \lambda E(x)$ or not. In this case $\lambda = 1/\log 2$. The first class goes as before, using Theorem 08 or Halász' original theorem. The upper bound for H_2 is more delicate. We write $n = ab$, with $P^+(a) \leqslant z/y < P^-(b)$, and use Theorem 07 to show that, with negligible exceptions, we have $a \leqslant (z/y)^{cE(x)}$ for a suitable constant c. There is a further split according as $\Omega(b, z) > \lambda E(x)$ or not. In the process of counting this last class we introduce the function $\tau(b; y/a, z)$: sums involving this can be estimated more precisely for small a, say $a \leqslant (z/y)^{c \log E(x)}$, and when $(z/y)^{c \log E(x)} < a \leqslant (z/y)^{cE(x)}$, effective use of the bounds for the Ψ-function, given in Theorem 05, is possible.

2.7 Proof of Theorem 21 (ii), (iii) – lower bounds

We give a detailed proof for part (ii) and indicate the modifications needed for part (iii).

Let $2 \leqslant y < z \leqslant \min(2y, \sqrt{x})$ and $\beta \leqslant \log 4 - 1 + O(1)/\sqrt{\log_2 y}$. We assume to begin with that $\beta \leqslant \log 4 - 1$; the other case will be dealt with very briefly.

If χ denotes the characteristic function of an arbitrary set of integers,

Cauchy's inequality gives

$$\left(\sum_{n\leqslant x}\chi(n)\tau(n;y,z)\right)^2 \leqslant H(x,y,z)\sum_{n\leqslant x}\chi(n)\tau(n;y,z)^2. \qquad (2.43)$$

This is the basis of our method – of course the critical step is a good choice for χ. Our idea is to impose the weakest condition which will ensure that for the dominant n in both the sums in (2.43) we have $\tau(n;y,z)\leqslant L(\log y)^c$. To this extent $\chi(n)\tau(n;y,z)$ will mimic the characteristic function of the set counted by $H(x,y,z)$.

We set $\lambda=(1+\beta)/\log 2\leqslant 2$. Simple heuristic arguments then suggest that we put $\chi(n)=\chi(n,\lambda)$, where $\chi(n,\lambda)=1$ if and only if

$$\Omega(n,t)\leqslant \lambda\log_2 3t+4\log(L(\log y)) \quad (2\leqslant t\leqslant y). \qquad (2.44)$$

We are going to show that, with this choice of χ, we have both

$$R:=\sum_{n\leqslant x}\chi(n)\tau(n;y,z)\gg x(\log y)^{-Q(\lambda)}L(\log y)^{-c_3}, \qquad (2.45)$$

and

$$S:=\sum_{n\leqslant x}\chi(n)\binom{\tau(n;y,z)}{2}\ll x(\log y)^{-Q(\lambda)}L(\log y)^{c_4}. \qquad (2.46)$$

We begin with (2.45). We have

$$R=\sum_{\substack{md\leqslant x\\ y<d\leqslant z}}\chi(md)=\sum_{\substack{md\leqslant x\\ y<d\leqslant z}}\chi(qd)$$

where $m=qr$, $P^+(q)\leqslant y<P^-(r)$. Certainly $\chi(qd)=1$ will follow from

$$\Omega(q,t),\Omega(d,t)\leqslant\frac{\lambda}{2}\log_2 3t+2\log(L(\log y)) \quad (t\leqslant y), \qquad (2.47)$$

and we construct such integers q, d in the following manner. Put $L=L(\log y)$, and define the intervals

$$I(j)=(\exp L^j,\exp L^{j+1}] \quad (j\geqslant 1).$$

We set

$$h=\left[\frac{\lambda}{2}\log L\right],\quad k=\left[\frac{\log_2 y}{\log L}\right],$$

and assume y sufficiently large that $k\geqslant 3$. Let a_j denote, generically, an integer equal to the product of h primes, distinct or not, from $I(j)$, and a denote an integer of the form

$$a=\prod_{1\leqslant j<k-1}a_j. \qquad (2.48)$$

We shall restrict q and d to be of the form $q=a$, $d=pa'$, $y<pa'\leqslant z$. It is then easy to check that (2.47) holds – we leave this as an exercise.

We need an upper bound for a. We have

$$a \leqslant \exp\left\{h \sum_{1 \leqslant j < k-1} L^{j+1}\right\} < \exp\{2hL^{k-1}\} < y^{2h/L} = y^{o(1)},$$

whence for large y, we have (since q is an a),

$$qd \leqslant y^{1/2} z \leqslant y^{-1/2} x. \tag{2.49}$$

We may now embark on the lower bound computation. From the above,

$$R \geqslant \sum_{\substack{qdr \leqslant x \\ y < d \leqslant z}} 1 = \sum_{\substack{qd \leqslant x \\ y < d \leqslant z}} \Phi\left(\frac{x}{qd}, y\right).$$

By (2.49) and Theorem 06, we have

$$\Phi\left(\frac{x}{qd}, y\right) \gg x/qd \log y$$

moreover the condition $qd \leqslant x$ is redundant. So

$$R \gg \frac{x}{\log y} \sum_{a} \frac{1}{a} \sum_{a'} \frac{1}{a'} \sum_{\substack{y/a' < p \\ \leqslant z/a'}} \frac{1}{p}.$$

By the Prime Number Theorem, the p sum is $\asymp (\log y)^{-1-\beta}$, and we obtain

$$R \gg \frac{x}{(\log y)^{2+\beta}} \prod_{1 \leqslant j < k-1} \left(\sum \frac{1}{a_j}\right)^2$$

$$\gg \frac{x}{(\log y)^{2+\beta}} \prod_{1 \leqslant j < k-1} \left(\frac{1}{h!}\left(\sum_{p \in I(j)} \frac{1}{p}\right)^h\right)^2.$$

Using the modest approximation

$$\sum_{p \in I(j)} \frac{1}{p} = \left(1 + O\left(\frac{1}{L}\right)\right) \log L$$

and Stirling's formula, we readily find that the product above is

$$\gg (\log y)^{-Q(\lambda) + 2 + \beta} L^{-c_3}$$

which yields (2.45). Notice that Stirling's formula gives rise to a factor $\sqrt{h}^{-2k}$, and it is to keep this in check that we need L so large.

It remains to deal with (2.46). We shall need convenient upper bounds for $\chi(n)$ and we fall back on the general principle of §0.5, writing

$$\chi(n) \leqslant v^{\Omega(n,t)} (\log 3t)^{-\lambda \log v} L^{-4 \log v} \tag{2.50}$$

where $v \in (0, 1]$ and $t \in [1, y]$ are parameters at our disposal. We shall also use

$$\binom{\tau(n; y, z)}{2} = \sum_{\substack{d, d' \mid n \\ y < d < d' \leqslant z}} 1 \leqslant \sum_{m \mid n} \sum_{\substack{tt' \mid n/m \\ y/m < t < t' \leqslant z/m}} 1 \tag{2.51}$$

and we bear in mind that, as in §2.5, the inner sum on the right is empty unless $m \leqslant \eta y := z - y$.

Let α be fixed, $0 < \alpha \leqslant 1$. From (2.50) and (2.51), we deduce that

$$S := \sum_{n \leqslant x} \chi(n) \binom{\tau(n; y, z)}{2} \leqslant L^{-4 \log \alpha}(S_1 + S_2) \tag{2.52}$$

where

$$S_1 := \sum_{n \leqslant x} v^{\Omega(n,y)} (\log 3y)^{-\lambda \log v} \sum_{\substack{m|n \\ m \leqslant \sqrt{y}}} w^{\Omega(n,m)} (\log 3m)^{-\lambda \log w} \sum_{\substack{tt'|n/m \\ y/m < t < t' \leqslant z/m}} 1,$$

$$S_2 := \sum_{n \leqslant x} v^{\Omega(n,y)} (\log 3y)^{-\lambda \log v} \sum_{\substack{m|n \\ \sqrt{y} < m \leqslant \eta y}} w^{\Omega(n,y/m)} \left(\log \frac{3y}{m} \right)^{-\lambda \log w} \sum_{y/m < t < t' \leqslant z/m} 1,$$

provided $v \leqslant 1$, $w \leqslant 1$, $vw \geqslant \alpha$. Notice that we have used the fact that $\chi(n) = \chi(n)^2$, applying (2.50) twice in each sum, and that v and w need not take the same values in S_1 and S_2.

In S_1 we write $n = mtt'd$ and sum over d first. This gives

$$S_1 = (\log 3y)^{-\lambda \log v} \sum_{m \leqslant \sqrt{y}} (vw)^{\Omega(m)} (\log 3m)^{-\lambda \log w}$$
$$\times \sum_{y/m < t < t' \leqslant z/m} v^{\Omega(tt',y)} w^{\Omega(tt',m)} \sum_{d \leqslant x/mtt'} v^{\Omega(d,y)} w^{\Omega(d,m)}.$$

We have $v^{\Omega(d,y)} \leqslant v^{\Omega(d,m)}$ and $x/mtt' \geqslant xm/z^2 \geqslant m$. Hence the inner sum is $\ll (x/mtt')(\log 3m)^{vw-1}$ by Theorem 01. Also $\Omega(tt', y) \geqslant \Omega(tt') - 2$, and so

$$S_1 \ll_\alpha x (\log 3y)^{-\lambda \log v} \sum_{m \leqslant \sqrt{y}} (vw)^{\Omega(m)} (\log 3m)^{vw - 1 - \lambda \log w} m^{-1} Z$$

where

$$Z = \sum_{y/m < t \leqslant z/m} \frac{1}{t} v^{\Omega(t)} w^{\Omega(t,m)} \ll (\log y)^{v - 1 - \beta} (\log 3m)^{vw - v}$$

by Shiu's theorem: this applies because $y/m \geqslant \sqrt{y}$. Thus

$$S_1 \ll_\alpha x (\log y)^{2v - 2 - 2\beta - \lambda \log v} \sum_{m \leqslant \sqrt{y}} \frac{1}{m} (vw)^{\Omega(m)} (\log 3m)^{3vw - 2v - 1 - \lambda \log w}.$$

The m-sum may now be majorized by partial summation using Theorem 01 again. It is

$$\ll (\log y)^{4vw - 2v - \lambda \log w - 1} \log_2 y$$

provided

$$4vw - 2v - \lambda \log w - 1 \geqslant 0 \tag{2.53}$$

(else it is the partial sum of a convergent series). Assuming this last condition is realized, we find that

$$S_1 \ll_\alpha x (\log y)^{4vw - \lambda \log vw - 2\beta - 3} \log_2 y$$

and the exponent takes its minimum value $-Q(\lambda)$ when $vw = \lambda/4$. We choose $v = \lambda/2$, $w = \frac{1}{2}$: the left-hand side of (2.53) then equals $\beta \geqslant 0$, and we have

$$S_1 \ll_\alpha x(\log y)^{-Q(\lambda)} \log_2 y. \tag{2.54}$$

We turn our attention to S_2. Arguing as above, we arrive at

$$S_2 \ll_\alpha x(\log y)^{v-1-\lambda \log v} \sum_{\sqrt{y}<m\leqslant \eta y} \frac{1}{m} v^{\Omega(m)} w^{\Omega(m,y/m)} \times \left(\log \frac{3y}{m}\right)^{vw-v-\lambda \log w} \left(\sum_{y/m<t\leqslant z/m} \frac{1}{t}(vw)^{\Omega(t)}\right)^2.$$

We cannot appeal to Shiu's theorem to bound the sum over t, because $(z-y)/m$ may be smaller than any power of y/m, indeed it may equal 1. We use (2.39) instead; we have $m \leqslant \eta y$ so

$$\sum_{y/m<t\leqslant z/m} \frac{(vw)^{\Omega(t)}}{t} \ll (\log y)^{-\beta} \left(\log \frac{3y}{m}\right)^{vw-1} \log_2 y,$$

and

$$S_2 \ll_\alpha x(\log y)^{v-1-2\beta-\lambda \log v} (\log_2 y)^2 \sum_{\sqrt{y}<m\leqslant \eta y} \frac{1}{m} v^{\Omega(m)} w^{\Omega(m,y/m)} \left(\log\left(\frac{3y}{m}\right)\right)^A$$

where

$$A = A(v, w) = 3vw - v - 2 - \lambda \log w.$$

Put

$$K(s) := \sum_{m\leqslant s} v^{\Omega(m)} w^{\Omega(m,y/m)}.$$

Since $\Omega(m, y/m) \geqslant \Omega(m, y/s)$ and $w \leqslant 1$, we have for $\sqrt{y} \leqslant s \leqslant y$

$$K(s) \leqslant \sum_{m\leqslant s} v^{\Omega(m)} w^{\Omega(m,y/s)} \ll s(\log s)^{v-1} \left(\log \frac{3y}{s}\right)^{vw-v}$$

by Theorem 01. Whence the inner sum above is

$$\int_{\sqrt{y}}^{\eta y} \frac{1}{s}\left(\log \frac{3y}{s}\right)^A dK(s) \ll (\log y)^{A+vw} \log_2 y$$

by partial integration, provided

$$A + vw - v = 4vw - 2v - 2 - \lambda \log w \geqslant -1. \tag{2.55}$$

We then have

$$S_2 \ll_\alpha x(\log y)^{4vw-\lambda \log vw - 2\beta - 3} (\log_2 y)^3.$$

The exponent is a minimum when $vw = \lambda/4$, and we select $v = \lambda/2$, $w = \frac{1}{2}$ as before, so that (2.55) is satisfied. This yields

$$S_2 \ll_\alpha x(\log y)^{-Q(\lambda)} (\log_2 y)^3$$

and we deduce from this, (2.52) and (2.54) that

$$S \ll x(\log y)^{-Q(\lambda)} L(\log y)^{12},$$

(taking $\alpha = 1/\log 16$). This proves (2.46).

We may now obtain the lower bound for $H(x, y, z)$. Notice that (2.43) may be re-written in the form

$$H(x, y, z) \geqslant R^2/(2S + R)$$

and that the right-hand side is an increasing function of R. We insert (2.45) and (2.46); this yields the lower bound in (2.10) provided $\beta \leqslant \log 4 - 1$, because $Q(\lambda) = G(\beta)$.

To complete the proof of part (ii), we have to consider the case $\beta > \log 4 - 1$, $\xi = O(1)$. Let β_1, ξ_1 and z_1 be related in the same way as β, ξ and z, that is by (2.8), and let $\xi_1 = \xi_1(y) \to \infty$ as $y \to \infty$. For large y, $\xi_1 > \xi$ and $z_1 < z$, so that $H(x, y, z) \geqslant H(x, y, z_1)$. By part (i) of the theorem, we have

$$\begin{aligned} H(x, y, z_1) &= (1 + o(1))x(\log y)^{-\beta_1} \\ &\gg x(\log y)^{-\beta} \exp\{(\xi - \xi_1)\sqrt{\log_2 y}\} \\ &\gg x(\log y)^{-G(\beta)} L(\log y)^{-c_1} \end{aligned}$$

provided $\xi_1 \leqslant c_1\sqrt{\log_3 y}$. The result follows.

The lower bound in (2.11), valid for the range

$$2y < z \leqslant \min(y^{3/2}, x^{1/2})$$

may be proved in a similar fashion, a slight extra complication arising from the 'small' primes $\leqslant z/y$. We have to consider, instead of $\tau(n; y, z)$ the quantity $\tau(b; y, z)$ where b is the largest divisor of n free of prime factors $\leqslant z/y$. The definition of χ has to be modified too: we ignore 'small' prime factors. There is a simplification in the computations inasmuch as $z - y \geqslant y$. Thus there are no 'short' sums of multiplicative functions.

We shall not give a complete proof here, hoping that the reader who has reached this point will be able to see the main lines of the argument and moreover cope with the rather similar technical difficulties which arise. For a detailed proof we refer him to Tenenbaum (1984).

2.8 Proof of Theorem 21 (iv) – large z

We prove here that the formula (2.12),

$$H(x, y, z) = x\left(1 + O\left(\frac{\log 2y}{\log z}\right)\right),$$

holds, uniformly for $1 \leqslant y < z \leqslant x$. Let P be the product of the primes

$p \in (y, z]$. Then

$$0 \leqslant [x] - H(x,y,z) \leqslant \sum_{\substack{n \leqslant x \\ (n,P)=1}} 1 \ll x \prod_{y<p\leqslant z}\left(1-\frac{1}{p}\right) \ll x\frac{\log 2y}{\log z}$$

by Theorem 01 and Mertens' Formula. This is all we need.

We remark that, although easy, the upper bound for $[x] - H(x,y,z)$ is essentially optimal. Notice that

$$[x] - H(x,y,z) \geqslant \sum_{m \leqslant y} \Phi\left(\frac{x}{m}, z\right)$$

$$\gg \frac{x}{\log z} \log \min\left(\frac{x}{2z}, y\right)$$

by Theorem 06. Suppose $a \in (0,1)$ fixed, $z \leqslant xy^{-a}$, $y > y_0(a)$. Then

$$[x] - H(x,y,z) \gg ax\frac{\log y}{\log z}.$$

We see that if $(\log z)/(\log y)$ is large it is almost the same to ask for a divisior as a prime factor in $(y, z]$.

Notes on Chapter 2

§2.1. Sets of multiples. Let $\mathscr{A}'$ be the intersection of all the sequences $\mathscr{A}$ such that $\mathscr{B}(\mathscr{A}) = \mathscr{B}$. It is clear that $\mathscr{A}'$ will be both minimal and primitive, provided $\mathscr{B}(\mathscr{A}') = \mathscr{B}$. It is sufficient to prove that for two sequences $\mathscr{A}_1, \mathscr{A}_2$ such that $\mathscr{B}(\mathscr{A}_1) = \mathscr{B}(\mathscr{A}_2) = \mathscr{B}$, we have $\mathscr{B}(\mathscr{A}_1 \cap \mathscr{A}_2) = \mathscr{B}$. Suppose not, and that b is the least member of $\mathscr{B} \backslash \mathscr{B}(\mathscr{A}_1 \cap \mathscr{A}_2)$. Then $b = m_1 a_1 = m_2 a_2$ where $a_1 \in \mathscr{A}_1, a_2 \in \mathscr{A}_2$ and plainly $a_1 \neq a_2$. Suppose for example $a_1 < a_2$. Then $m_1 > 1$. Also $a_1 \in \mathscr{B}(\mathscr{A}_1) = \mathscr{B}$, and since $a_1 < b$, $a_1 \in \mathscr{B}(\mathscr{A}_1 \cap \mathscr{A}_2)$, that is $a_1 = m_3 a_3$ where $a_3 \in \mathscr{A}_1 \cap \mathscr{A}_2$. But then $b = (m_1 m_3) a_3$ and this is a contradiction.

§2.2. Apart from the problem of an asymptotic formula for $H(x, y, z)$ the main open question in this field of investigation is that of the conditional probability that $\tau(n; y, z) = k$ given $\tau(n; y, z) \neq 0$. Define

$$H_k(x, y, z) := \operatorname{card}\{n : n \leqslant x, \tau(n; y, z) = k\}$$

$$\varepsilon_k(y) := \lim_{x \to \infty} x^{-1} H_k(x, y, 2y), \quad \varepsilon(y) = \sum_{k=1}^{\infty} \varepsilon_k(y).$$

Erdős has conjectured for some time that

$$\lim_{y \to \infty} \frac{\varepsilon_k(y)}{\varepsilon(y)} = 0; \tag{1}$$

indeed at one time this was considered by him as a possible strategy for his conjecture dealt with in Chapter 5. At present (1) is unsettled, but we think it may be false. Recent work of Tenenbaum (1988) shows that $\varepsilon_1(y)/\varepsilon(y) = (\log y)^{o(1)}$, more precisely (with L as in §2.2),

$$L(\log y)^{-c} \ll \varepsilon_1(y)/\varepsilon(y) \leqslant 1 \quad (y \geqslant 2).$$

In the same paper, estimates comparable to those of Theorem 21 are given for $H_k(x, y, z)$, $k \in \mathbb{Z}^+$ fixed. In view of these results it is tempting to formulate the following conjecture.

Conjecture. *For each* $k \geqslant 1$, *there exists* $\lim_{y \to \infty} (\varepsilon_k(y)/\varepsilon(y)) =: d_k$ *with* $d_k > 0$ *and* $\sum d_k = 1$.

Erdős has also conjectured that $\varepsilon_1(y, z) := \lim_{x \to \infty} x^{-1} H_1(x, y, z)$ is a unimodular function of z. Precisely where $\varepsilon_1(y, z)$ reaches its maximum is then a very interesting question. From Tenenbaum's results, it follows that $\varepsilon_1(y, z)$ is unimodular to within a factor of incertitude $L(\log y)^{O(1)}$.

Exercises on Chapter 2

20. Put $d(y)=\lim_{x\to\infty} x^{-1}H(x,y,2y)$ and suppose that (2.1) holds, i.e. $\liminf d(y)=0$. Show that given $\varepsilon\in(0,\frac{1}{2})$ there exists $\{y_k\to\infty\}$ such that

(i) $d(y_k)\leqslant\varepsilon\cdot 2^{-k-1}\quad(k\geqslant 0)$,

(ii) $H(x,y_k,2y_k)\leqslant 2xd(y_k)\quad(k\geqslant 0, x\geqslant y_{k+1})$.

Now set $\mathscr{A}:=\bigcup_k(y_k,2y_k]\cap\mathbb{Z}^+$ and let $B(x)$ denote the counting function of $\mathscr{B}(\mathscr{A})$. Show that

(iii) $\limsup x^{-1}B(x)\geqslant\frac{1}{2}$,

(iv) $\liminf x^{-1}B(x)\leqslant\varepsilon$.

Deduce from Theorem 21 that, for sufficiently large c, we may take $y_k=\exp\{c(2^k/\varepsilon)^{1/\delta}\}$, δ as in (2.3).

21. For $\lambda>0$ set $\mathscr{A}_\lambda:=\bigcup_k(\exp k^\lambda, 2\exp k^\lambda]\cap\mathbb{Z}^+$ and $\mathscr{B}_\lambda=\mathscr{B}(\mathscr{A}_\lambda)$. Prove that $\lambda>1/\delta\Rightarrow\bar{\mathbf{d}}\mathscr{B}_\lambda<1$. [Hint: Use Behrend's inequality (1948): *if* $T(a_1,a_2,\ldots,a_m)$ *denotes the density of the integers divisible by no* a_i, *then for any* a_i, b_j

$$T(a_1,\ldots,a_m,b_1,\ldots,b_n)\geqslant T(a_1,\ldots,a_m)T(b_1,\ldots,b_n). \tag{1}$$

(For alternative proofs of (1) see Halberstam & Roth (1966), Chapter 5, or Ruzsa (1976).) Hall & Tenenbaum (1986) prove that $\mathbf{d}\mathscr{B}_\lambda=1$ for $\lambda<\lambda_0=1.314\,57\ldots$, and conjecture that $\mathbf{d}\mathscr{B}_\lambda=1\Leftrightarrow\lambda<\lambda_1=1/(1-\log 2)$.]

22. Put $\Delta(n,t):=\tau(n;t,et)$, $\Delta(n)=\max\Delta(n,t)$ and $\varepsilon(n,t)=0$ or 1 according as $\Delta(n,t)=0$ or not. Show that, uniformly for $x\geqslant t\geqslant 1$, both

(i) $\sum_{n\leqslant x}\Delta(n,t)=x+O(x/t)$

and

(ii) $\sum_{n\leqslant x}\Delta(n,t)^2\ll x\log 2t$.

Now let $c>\log(1/\delta)$ (δ as in (2.3)), and for $1\leqslant j\leqslant k=[(\log_3 x)/2c]$ set $t_j=\exp\exp\exp(cj)$. Set $j(n)=\max\{j: 1\leqslant j\leqslant k, \Delta(n,t_j)\neq 0\}$. Show that

(iii) $\sum_{n\leqslant x}\Delta(n,t_{j(n)})\geqslant\sum_{1\leqslant j<k}\sum_{n\leqslant x}\Delta(n,t_j)\Bigl(1-\sum_{j<m\leqslant k}\varepsilon(n,t_m)\Bigr)$,

and for all j, m, $1\leqslant j<m\leqslant k$, that

(iv) $\sum_{n\leqslant x}\Delta(n,t_j)\varepsilon(n,t_m)\ll x\exp\{-\frac{1}{2}(\delta-e^{-c})e^{cm}\}$.

Deduce that

(v) $\sum_{n\leqslant x}\Delta(n)\geqslant x(k+O(\log k))$.

23. Let $\theta\in C^2[0,1]$ be real, $F(n;\theta):=\sum_{i<\tau(n)}\theta(d_i/d_{i+1})$, where $d_i=d_i(n)$. Put $k=[\tau(n)/2]$. Show that

(i) $F(n;\theta)=2\sum_{1\leqslant i<k}\theta(d_i/d_{i+1})+O(1)$

Put $\varphi(t)=\theta'(t)+t\theta''(t)$, $\varepsilon(n;y,z)=1$ or 0 according as n has a divisor in $(y,z]$ or not. Prove

(ii) $\displaystyle\int_1^{d_k}\int_1^z\varphi\left(\frac{y}{z}\right)\varepsilon(n;y,z)\frac{dy\,dz}{z^2}=(k-2)\theta(1)-\sum_{i<k}\theta\left(\frac{d_i}{d_{i+1}}\right)+\theta\left(\frac{1}{d_k}\right)$

(iii) $\displaystyle\int_{d_k}^{\sqrt{x}}\int_1^z\varphi\left(\frac{y}{z}\right)\varepsilon(n;y,z)\frac{dy\,dz}{z^2}\ll\log(2x/n)\quad(n\leqslant x)$.

Hence obtain the formula (2.5).

24. With the same notation as in Exercise 23, prove that

$$\sum_{n\leqslant x}F(n;\theta)=\left(\theta(1)+O\left(\frac{\log_3 x}{(\log x)^\delta\sqrt{\log_2 x}}\right)\right)x\log x,$$

and show that the exponent δ is best possible.

25. Let $\varepsilon\in(0,1)$ be fixed, and define

$$\mathrm{Prop}(n)=\mathrm{card}\{i:1\leqslant i<\tau(n),\,d_{i+1}(n)<(1+\varepsilon)d_i(n)\}.$$

Use Exercise 24 to establish

$$\sum_{n\leqslant x}\mathrm{Prop}(n)=\left(1+O\left(\frac{\log_3 x}{(\log x)^\delta\sqrt{\log_2 x}}\right)\right)x\log x.$$

26. Define

$$k_r(n):=\sum_{1\leqslant i<\tau(n)}(d_i(n)/d_{i+1}(n))^r.$$

Prove that

$$\sum_{n\leqslant x}k_r(n)=(1-K_r(x)/(\log x)^\delta)x\log x$$

with $e^{-\varepsilon r}L(\log x)^{-c}\ll_\varepsilon K_r(x)\ll r\log_3 x/\sqrt{\log_2 x}$, uniformly for $r\geqslant r_0>0$. [Hint: use (2.5).]

27. Estimate

$$\sum_{n\leqslant x}\sum_{1\leqslant i<\tau(n)}\left(1-\frac{d_i(n)}{d_{i+1}(n)}\right)^2.$$

28. Let $\tau^+(n) := \sum \varepsilon(n; 2^k, 2^{k+1})$, that is the number of k for which $(2^k, 2^{k+1}]$ contains at least one divisor of n.

Prove that $\tau^+(n) \leqslant \min(\tau(n), [\log n/\log 2] + 1)$ and deduce by the method of §0.5 that

$$\sum_{n \leqslant x} \tau^+(n) \ll x(\log x)^{1-\delta}.$$

Use Theorem 21 to divide the right-hand side by $\sqrt{\log_2 x}$.

29. Let $g(n) := \text{card}\{i: 1 \leqslant i < \tau(n),\ d_i(n) | d_{i+1}(n)\}$. Find an inequality between $g(n)$ and $\tau^+(n)$ and deduce an upper bound for $\sum\{g(n): n \leqslant x\}$. [This sum is $\gg x(\log x)^{1-\delta} L(\log x)^{-c}$ (cf. Tenenbaum (1988)).]

210. For $x \geqslant z \geqslant y \geqslant t \geqslant 1$ set $H(x, y, z; t) = \sum_{n \leqslant x, P^-(n) > t} \varepsilon(n; y, z)$. Show that, for $v \in (0, 1]$, $w \geqslant 1$, an upper bound for $H(x; y, z; t)$ is

$$\left(\frac{\log y}{\log t}\right)^{-\log v/\log 2} \sum_{\substack{n \leqslant x \\ P^-(n) > t}} v^{\Omega(n,y)} \tau(n; y, z) + \left(\frac{\log y}{\log t}\right)^{-\log w/\log 2} \sum_{\substack{n \leqslant x \\ P^-(n) > t}} w^{\Omega(n,y)}.$$

Hence show that, for $2y \leqslant z \leqslant \min(y^2, \sqrt{x})$,

$$H(x, y, z; t) \ll x(\log 2y)^{-\delta}(\log 2t)^{\delta-2} \log\left(\frac{tz}{y}\right).$$

211. Let ε_y denote the density of the integers n for which

$$M(n, y) = \sum_{\substack{d|n \\ d \leqslant y}} \mu(d) \neq 0.$$

Prove that $\varepsilon_y \ll (\log y)^{-\delta}$ by first establishing

$$\varepsilon_y \leqslant \lim_{x \to \infty} x^{-1} \sum_{p \leqslant \sqrt{y}} H\left(\frac{x}{p}, \frac{y}{p}, y; p - 1\right) + O(1/\log y),$$

and using the result of Exercise 210.

3
Imaginary powers

3.1 Introduction

For each n and real θ we define

$$\tau(n,\theta)=\sum_{d\mid n} d^{i\theta} \tag{3.1}$$

and we have

$$\sum_{\substack{d\mid n\\ a\leqslant \log d\leqslant b}}^{*} 1=\frac{1}{2\pi}\int_{-\infty}^{\infty}\tau(n,\theta)\frac{\mathrm{e}^{-ib\theta}-\mathrm{e}^{-ia\theta}}{-\mathrm{i}\theta}\,\mathrm{d}\theta \tag{3.2}$$

(where the * denotes that if e^a or $\mathrm{e}^b|n$, their contribution to the sum on the left is $\frac{1}{2}$). Thus in principle any question about the divisors of n can be re-stated in terms of $\tau(n,\theta)$, and this is a useful idea because $\tau(n,\theta)$ is multiplicative. It makes an appearance in some form in each of the next three chapters of this book.

If $|\tau(n,\theta)|$ is appreciably smaller than $\tau(n)$, and this for a suitable range of values of θ, we may infer some degree of uniformity in the distribution of the divisors of n. Thus the information that is usually needed is an upper bound for a suitable average, possibly over both n and θ, of $|\tau(n,\theta)|$.

For fixed non-zero θ, the function $|\tau(n,\theta)|$ has average order $c(\theta)\log^a n$ ($a=4/\pi-1=0.273\,23\ldots$), whereas the average order of $\tau(n)$ is $\log n$. What is perhaps surprising is that, for almost all n, $\tau(n,\theta)\ll\tau(n)^{\varepsilon}$, and we begin by proving this. We avoid the phrase 'normal order' which has a precise technical meaning, inapplicable here (cf. §4.4 for a discussion of this point).

3.2 A p.p. upper bound for $|\tau(n,\theta)|$

Theorem 30. *Let θ be fixed, real, and non-zero, and $\xi(n)\to\infty$ as $n\to\infty$. Then*

$$|\tau(n,\theta)|<\xi(n)^{\sqrt{\log\log n}}\quad \text{p.p.} \tag{3.3}$$

The function $\tau(n,\theta)$ oscillates rapidly as θ varies, vanishing wherever $\tau(p^{\alpha},\theta)$ vanishes for every p, $p^{\alpha}\,\|\,n$. Because the numbers $\log p$ are independent over the rationals these zeros are simple, and there are about $(2\pi)^{-1}(b-a)\log n$ of them on the interval $a\leqslant\theta\leqslant b$.

A lower bound complementing (3.3) would require arithmetic conditions on θ. Another open problem is to determine a p.p. upper bound for

$$\max_{a\leqslant\theta\leqslant b}|\tau(n,\theta)| \tag{3.4}$$

where $a<b, 0<ab$. What is the infimum of the numbers μ such that this is $\ll_\mu(\log n)^\mu$ p.p.?

The proof of (3.3) depends on the *method of low moments.* The name will become clear as we proceed. We need some preliminary results which will also be useful in later chapters.

Lemma 30.1. *Let f be a periodic function of bounded variation over the period $[0,2\pi]$ and having mean value*

$$\bar{f}:=\frac{1}{2\pi}\int_0^{2\pi}f(v)\,\mathrm{d}v.$$

Then, for real θ, w, z such that $\theta\neq 0$, $1<w<z$, and every positive c, we have

$$\sum_{w<p\leqslant z}\frac{1}{p}f(\theta\log p)=\bar{f}\log\left(\frac{\log z}{\log w}\right)+O\left(\frac{V(f)}{|\theta|\log w}\right)$$
$$+O_c((M(f)+(1+|\theta|)V(f))\mathrm{e}^{-c\sqrt{\log w}}) \tag{3.5}$$

where

$$M(f):=\sup_v|f(v)|,\quad V(f):=\int_0^{2\pi}|\mathrm{d}f(v)|.$$

Proof. We may assume $\theta>0$. We require the Prime Number Theorem in the form $\mathscr{E}(t):=\pi(t)-\operatorname{li}t\ll_c t\exp(-2c\sqrt{\log t})$. Partial summation yields

$$\sum_{w<p\leqslant z}\frac{1}{p}f(\theta\log p)=\int_w^z\frac{f(\theta\log t)}{t\log t}\,\mathrm{d}t$$
$$+\left[\frac{\mathscr{E}(t)}{t}f(\theta\log t)\right]_w^z-\int_w^z\mathscr{E}(t)\mathrm{d}\left(\frac{1}{t}f(\theta\log t)\right)$$
$$=\bar{f}\log\left(\frac{\log z}{\log w}\right)+\int_{\theta\log w}^{\theta\log z}(f(v)-\bar{f})\frac{\mathrm{d}v}{v}$$
$$-\int_w^z\frac{\mathscr{E}(t)}{t}\mathrm{d}f(\theta\log t)+O_c(M(f)\mathrm{e}^{-c\sqrt{\log w}}).$$

The second term here is $\ll V(f)/|\theta|\log w$, applying the second mean value theorem to the real and imaginary parts, and noticing that for all a, b

$$\left|\int_a^b(f(v)-\bar{f})\,\mathrm{d}v\right|\leqslant\frac{\pi}{4}V(f). \tag{3.6}$$

The third term is (using the periodicity of f),

$$\ll_c e^{-c\sqrt{\log w}} \int_{\theta \log w}^{\theta \log w + 2\pi} |\mathrm{d} f(v)| \sum_{k=0}^{\infty} \exp\left(-c\sqrt{\left(\frac{v+2\pi k}{\theta}\right)}\right)$$
$$\ll_c (1+|\theta|) V(f) e^{-c\sqrt{\log w}}$$

since the sum over k is $\ll_c (1+|\theta|)$ uniformly for $v \geqslant 0$. This proves (3.5).

Lemma 30.2. *Uniformly for $\lambda > 0$ and real θ, $|\theta| \geqslant 1/\log x$, we have, for every positive c,*

$$\sum_{p \leqslant x} \frac{1}{p} |\tau(p,\theta)|^\lambda = F(\lambda) \log_2 x + (2^\lambda - F(\lambda)) \log^+ \frac{1}{|\theta|}$$
$$+ A(\lambda,\theta) + O\left(\frac{2^\lambda}{|\theta| \log x}\right) + O_c((1+|\theta|)2^\lambda e^{-c\sqrt{\log x}}), \quad (3.7)$$

where $A(\lambda,\theta) \ll 2^\lambda(\lambda + \log_2(3+|\theta|))$, and

$$F(\lambda) := \frac{1}{2\pi} \int_0^{2\pi} |1+e^{iv}|^\lambda \, \mathrm{d}v = \frac{2^\lambda \Gamma(\frac{1}{2}+\frac{1}{2}\lambda)}{\sqrt{\pi} \cdot \Gamma(1+\frac{1}{2}\lambda)}. \quad (3.8)$$

Explicitly

$$A(\lambda,\theta) = F(\lambda)\gamma + \sum_p \left\{\frac{1}{p}|\tau(p,\theta)|^\lambda + F(\lambda) \log\left(1-\frac{1}{p}\right)\right\}$$
$$- (2^\lambda - F(\lambda)) \log^+ \frac{1}{|\theta|},$$

where γ is Euler's constant. (Another formula for $A(\lambda,\theta)$ appears in Exercise 34.)

Proof. We apply Lemma 30.1 with $f(v) = |1+e^{iv}|^\lambda$, $\bar{f} = F(\lambda)$, $M(f) = 2^\lambda$, $V(f) = 2M(f)$. Hence

$$\sum_{w < p \leqslant z} \frac{1}{p} |\tau(p,\theta)|^\lambda - F(\lambda) \log\left(\frac{\log z}{\log w}\right)$$
$$= O\left(\frac{2^\lambda}{|\theta| \log w}\right) + O_c((1+|\theta|)2^\lambda e^{-c\sqrt{\log w}}). \quad (3.9)$$

The right-hand side tends to zero as $w \to \infty$, whence, by Cauchy's criterion for convergence, there exists a constant $A' = A'(\lambda,\theta)$ such that

$$\sum_{p \leqslant z} \frac{1}{p} |\tau(p,\theta)|^\lambda = F(\lambda) \log_2 z + A'(\lambda,\theta) + \varepsilon(z;\lambda,\theta) \quad (3.10)$$

say, where $\varepsilon(z;\lambda,\theta) = o(1)\,(z \to \infty)$. Put $w = x$ in (3.9) and subtract the result from (3.10), then let $z \to \infty$. This yields

$$\sum_{p \leqslant x} \frac{1}{p} |\tau(p,\theta)|^{\lambda} = F(\lambda) \log_2 x + A'(\lambda,\theta)$$
$$+ \mathrm{O}\left(\frac{2^{\lambda}}{|\theta| \log x}\right) + \mathrm{O}_c((1+|\theta|)2^{\lambda} e^{-c\sqrt{\log x}}), \qquad (3.11)$$

and it remains to find a uniform estimate for $A'(\lambda,\theta)$.

If $|\theta| > 1$, we substitute $x = \exp(\log^2(3+|\theta|))$ in (3.11), taking $c = 1$, to deduce that $A'(\lambda,\theta) \ll 2^{\lambda} \log_2(3+|\theta|)$. If $|\theta| \leqslant 1$, we put $x = \exp(1/|\theta|)$. For $p \leqslant x$ we have $|\theta| \log p \leqslant 1$, so that $|\tau(p,\theta)| = 2\cos(\frac{1}{2}\theta \log p) > 2\cos(\pi/6)$. By the mean value theorem,

$$1 - \cos^{\lambda}\left(\frac{\theta}{2}\log p\right) < \frac{\lambda}{4} \sec\frac{\pi}{6} \cdot \theta^2 \log^2 p$$

whence

$$\sum_{p \leqslant x} \frac{1}{p} |\tau(p,\theta)|^{\lambda} = 2^{\lambda} \sum_{p \leqslant x} \frac{1}{p} + \mathrm{O}\left(\lambda \cdot 2^{\lambda} \theta^2 \sum_{p \leqslant x} \frac{\log^2 p}{p}\right)$$
$$= 2^{\lambda} \log \frac{1}{|\theta|} + \mathrm{O}((\lambda+1) \cdot 2^{\lambda}).$$

Comparing this with (3.11) we see that in this case $A'(\lambda,\theta) = (2^{\lambda} - F(\lambda)) \log(1/|\theta|) + \mathrm{O}((1+\lambda)2^{\lambda})$, and so for all θ, $\theta \neq 0$, we have

$$A'(\lambda,\theta) = (2^{\lambda} - F(\lambda)) \log^{+} \frac{1}{|\theta|} + \mathrm{O}(2^{\lambda}(\lambda + \log_2(3+|\theta|)))$$

and we set

$$A(\lambda,\theta) = A'(\lambda,\theta) - (2^{\lambda} - F(\lambda)) \log^{+} \frac{1}{|\theta|}$$

to obtain (3.7). For the explicit formula for $A(\lambda,\theta)$ we recall (Hardy & Wright, (1938), Theorem 428) that

$$\sum_{p \leqslant x} \frac{1}{p} = \log_2 x + \gamma + \sum_{p} \left\{ \log\left(1 - \frac{1}{p}\right) + \frac{1}{p} \right\} + \mathrm{o}(1),$$

and we multiply this by $F(\lambda)$, subtract it from (3.7), and let $x \to \infty$.

Lemma 30.3. *Uniformly for* $0 < y \leqslant y_0, 0 < \lambda \leqslant \lambda_0$, *and real* θ, $1/\log x \leqslant |\theta| \leqslant \exp(\sqrt{(\log x)})$, *we have*

$$\sum_{n \leqslant x} |\tau(n,\theta)|^{\lambda} y^{\omega(n)} \ll_{y_0,\lambda_0} x(\log x)^{F(\lambda)y-1}$$
$$\times \left(1 + \frac{1}{|\theta|}\right)^{2^{\lambda}y - F(\lambda)y} \log^{B}(3+|\theta|) \qquad (3.12)$$

where $B = B(y_0, \lambda_0)$.

Proof. We apply first Theorem 01 to deduce that the sum above is

$$\ll \frac{x}{\log x}\prod_{p\leqslant x}\left(1+\frac{y}{p}|\tau(p,\theta)|^{\lambda}+\cdots\right)\ll \frac{x}{\log x}\exp\left\{y\sum_{p\leqslant x}\frac{1}{p}|\tau(p,\theta)|^{\lambda}\right\}$$

and then apply Lemma 30.2.

Proof of Theorem 30. Since $\theta\neq 0$, we may assume that x is so large that $|\theta|\geqslant 1/\log x$. We fix θ, and put $y=y_0=\lambda_0=1$ in Lemma 30.3 so that we have, uniformly for $0<\lambda\leqslant 1$,

$$\sum_{n\leqslant x}|\tau(n,\theta)|^{\lambda}\ll_{\theta} x(\log x)^{F(\lambda)-1} \tag{3.13}$$

We have $F(0)=1$, and

$$F'(0)=\frac{1}{2\pi}\int_0^{2\pi}\log|1+e^{iv}|\,dv=0.$$

Moreover from (3.8), $F(\lambda)$ is analytic except for poles at $\lambda=-1,-3,-5,\ldots$, certainly $F\in C^2[0,1]$. Taylor's theorem therefore implies that, for a suitable constant κ, $F(\lambda)\leqslant 1+\kappa\lambda^2$, $0\leqslant\lambda\leqslant 1$.

We put $\lambda=1/\sqrt{\log_2 x}$ so that the right-hand side of (3.13) is $O(x)$. Now let $\xi(n)\to\infty$ as $n\to\infty$, and let E be the sequence of integers n such that

$$|\tau(n,\theta)|\geqslant \xi(n)^{\sqrt{\log\log n}}.$$

We put $\xi_1(x)=\min\{\xi(n):n>\sqrt{x}\}$. Then if $n\in E\cap(\sqrt{x},x]$ and x is sufficiently large, we have $|\tau(n,\theta)|^{\lambda}\geqslant\xi_1(x)^{1/2}$ whence

$$\operatorname{card}\{E\cap[1,x]\}\leqslant\sqrt{x}+\xi_1(x)^{-1/2}\sum_{n\leqslant x}|\tau(n,\theta)|^{\lambda}=o(x),$$

and E has asymptotic density 0 as required.

3.3 Ratios of divisors

Many questions concerning divisors, in particular some of the questions about propinquity, are better stated in terms of the ratios $r=d/d'$ $(d,d'|n)$. In some cases it is appropriate to count only distinct ratios: if $n=p_1^{\alpha_1}p_2^{\alpha_2}\ldots p_k^{\alpha_k}$ then these are the rational numbers

$$r=p_1^{\beta_1}p_2^{\beta_2}\ldots p_k^{\beta_k}\quad(-\alpha_i\leqslant\beta_i\leqslant\alpha_i,1\leqslant i\leqslant k), \tag{3.14}$$

and their number is

$$U(n)=\prod_{i=1}^{k}(2\alpha_i+1). \tag{3.15}$$

Notice that, for every n, we have

$$3^{\omega(n)}\leqslant U(n)\leqslant 3^{\Omega(n)}; \tag{3.16}$$

this 3 will make its appearance in several results in Chapters 4 and 5.

We label the distinct ratios in increasing order, $r_1 < r_2 < \cdots < r_U$, and we define

$$\rho(n, \theta) = \sum_{u=1}^{U} r_u^{i\theta} = \sum_{\substack{d,d' \mid n \\ (d,d')=1}} (d/d')^{i\theta}$$

$$= \prod_{p^\alpha \| n} (1 + 2\cos(\theta \log p) + \cdots + 2\cos(\theta\alpha \log p)). \tag{3.17}$$

This is analogous to the definition (3.1) of $\tau(n, \theta)$, and the two functions are connected by the relation

$$\rho(n, \theta) = \sum_{t \mid n} \mu(t) \left| \tau\left(\frac{n}{t}, \theta\right) \right|^2. \tag{3.18}$$

Of course if we count the ratios according to the number of representations $r = d/d'$, the function corresponding to $\rho(n, \theta)$ is just $|\tau(n, \theta)|^2$. There are analogues to the results of §3.2: we state without proof the following.

Theorem 31. *Let θ be fixed, real, and non-zero, and $\xi(n) \to \infty$ as $n \to \infty$. Then*

$$|\rho(n, \theta)| < \xi(n)^{\sqrt{\log\log n}} \quad \text{p.p.}$$

3.4 Average orders

We consider the average order of $|\tau(n, \theta)|^\lambda$ $(\lambda > 0)$, weighted with $y^{\omega(n)}$ or when appropriate $y^{\Omega(n)}$.

Theorem 32. *Let $y > 0$, $\lambda > 0$, $\theta \neq 0$. Then*

$$\sum_{n \leqslant x} |\tau(n, \theta)|^\lambda y^{\omega(n)} \sim \frac{x(\log x)^{F(\lambda)y - 1}}{\Gamma(F(\lambda)y)} \times \prod_p \left(1 - \frac{1}{p}\right)^{F(\lambda)y} \left(1 + \frac{y}{p}|\tau(p, \theta)|^\lambda + \cdots\right) \tag{3.19}$$

where $F(\lambda)$ is as defined in (3.8).

This is implied by a general theorem of Wirsing (1967) on sums of multiplicative functions together with Lemma 30.2. For the applications in Chapter 4, we need the case $\lambda = 2$, and a variant in which $|\tau(n, \theta)|^2$ is replaced by $\rho(n, \theta)$. We require a uniform error term. This can be quite modest and we avoid a more complicated formula and proof. We shall have $y \leqslant 1$ and it is convenient, though inessential, to replace ω by Ω.

Theorem 33. *Uniformly for real non-zero θ we have*

$$\sum_{n\leqslant x} |\tau(n,\theta)|^2 = \frac{6}{\pi^2}|\zeta(1+i\theta)|^2(x\log x + (2\gamma-1)x) + \{A(\theta) + \mathrm{Re}(A_2(\theta,1)x^{i\theta})\}x + O\{x^{1/2}(|\theta|^{1/3} + \log^{5/2} x)\log^{7/2} x\}, \tag{3.20}$$

where

$$A(\theta) = \frac{6}{\pi^2}\{\zeta'(1+i\theta)\zeta(1-i\theta) + \zeta(1+i\theta)\zeta'(1-i\theta) - 2\frac{\zeta'(2)}{\zeta(2)}|\zeta(1+i\theta)|^2\},$$

and, for $k = 1, 2, \theta \neq 0$, $|y| < 2$,

$$A_k(\theta, y) := 2\frac{\zeta^k(1+i\theta, y)\zeta(1+2i\theta, y)}{(1+i\theta)\zeta(2+2i\theta, y^2)}.$$

For each fixed y, $0 < y < 1$, and uniformly for real $\theta, 1/\log x \leqslant |\theta| \leqslant \exp\sqrt{\log x}$, we have

$$\sum_{n\leqslant x} |\tau(n,\theta)|^2 y^{\Omega(n)} = \frac{H^2(y)|\zeta(1+i\theta,y)|^2}{\Gamma(2y)\zeta(2,y^2)} x\log^{2y-1} x + \frac{H(y)}{\Gamma(y)}\{\mathrm{Re}(A_2(\theta,y)x^{i\theta})\}x\log^{y-1} x + O_y\{(|\theta|^{-1-2y} + \log^{13y+2}(3+|\theta|))x\log^{2y-2} x\} \tag{3.21}$$

and also

$$\sum_{n\leqslant x} \rho(n,\theta)y^{\Omega(n)} = \frac{H(y)}{\Gamma(y)}\left\{\frac{|\zeta(1+i\theta,y)|^2}{\zeta(2,y^2)} + \mathrm{Re}(A_1(\theta,y)x^{i\theta})\right\}x\log^{y-1} x + O_y\{(|\theta|^{-1-2y} + \log^{10y+2}(3+|\theta|))x\log^{y-2} x\} \tag{3.22}$$

where

$$\zeta(s,y) := \prod_p \left(1 - \frac{y}{p^s}\right)^{-1} = \sum_{n=1}^{\infty} \frac{y^{\Omega(n)}}{n^s}, \tag{3.23}$$

$$H(y) = \prod_p \left(1 - \frac{y}{p}\right)^{-1}\left(1 - \frac{1}{p}\right)^y. \tag{3.24}$$

The function $\zeta(s,y)$ is initially defined in the region $\sigma = \mathrm{Re}\, s > 1$, $y \leqslant 2$. The product formula extends the range of definition to $y > 2$, but there are then poles on the lines $\sigma = \log y/\log p$, for each $p < y$. We are interested in analytic continuation across the line $\sigma = 1$. The function

$$H(s,y) := \exp\left\{\sum_p \sum_m \frac{y^m - y}{p^{ms}}\right\} \tag{3.25}$$

is analytic in the half-plane $\sigma > \max(\frac{1}{2}, \log y/\log 2)$, and, in any simply connected sub-region of this half-plane which does not contain zeros of $\zeta(s)$ or the point $s=1$, we may define

$$\zeta(s,y) = H(s,y)\zeta(s)^y \tag{3.26}$$

as a single-valued, analytic function of s. We note that $H(y) = H(1,y)$.

Notice the differences between (3.21) and (3.22). The function $\rho(n,\theta)$ is not of fixed sign and there is some cancellation. The main term in (3.22) is smaller, and oscillates.

The substitution of Ω for the more usual ω in Theorem 33 enables us to employ a generalization of a formula of Ramanujan (1915),

$$\sum_{n=1}^{\infty} |\tau(n,\theta)|^2 \frac{y^{\Omega(n)}}{n^s} = \frac{\zeta^2(s,y)\zeta(s+i\theta,y)\zeta(s-i\theta,y)}{\zeta(2s,y^2)} \tag{3.27}$$

but is unimportant – there is an analogue of the theorem, with ω throughout, identical error terms, and slightly clumsier main terms.

Proof of Theorem 33. We denote the function (3.27) by $R(s)$ and employ Perron's formula in the form given by Titchmarsh (1951, Lemma 3.12). We make x half an odd integer (which does not affect the final result), $c = 1 + 1/\log x$, and we note that $|\tau(n,\theta)| \leqslant \tau(n) < n^{1/\log\log n}$ $(n > n_0)$ (Hardy & Wright (1938), Theorem 317). We choose $T = x + |\theta|$ if $y = 1$, $T = x^{4/\log\log x}$ else, and we have

$$\sum_{n \leqslant x} |\tau(n,\theta)|^2 y^{\Omega(n)} = \frac{1}{2\pi i}\int_{c-iT}^{c+iT} R(s)\frac{x^s}{s}\,ds + \begin{cases} O_\varepsilon(x^\varepsilon) & (y=1), \\ O(x^{1-1/\log\log x}) & (y<1). \end{cases}$$

When $y = 1$ we follow Titchmarsh (1951, §12.3 in the case $k = 4$, in which Titchmarsh has $\zeta^4(s)$ instead of $R(s)$), moving the line of integration across the poles to $\sigma = \frac{1}{2}$. The main terms in (3.20) come from the residues. On $\sigma = \frac{1}{2}$ we have $1/\zeta(2s) \ll \log T$, and we apply the Cauchy–Schwarz inequality twice to obtain

$$\ll (x^{1/2}\log T) J(0)^{1/2} J(\theta)^{1/4} J(-\theta)^{1/4}$$

where

$$J(\theta) := \int_{-T}^{T} |\zeta(\tfrac{1}{2} + i\theta + it)|^4 \frac{dt}{\sqrt{(t^2+\frac{1}{4})}} \ll |\theta|^{2/3} + \log^5 T.$$

This last inequality may be derived from the fourth power moment of $|\zeta(\frac{1}{2}+iu)|$ (Titchmarsh (1951), eq. (7.6.1)), combined with any upper bound $\zeta(\frac{1}{2}+iu) \ll |u|^\eta$ with $\eta < \frac{1}{6}$ (Titchmarsh (1951) §5.18). We obtain (3.20).

Now let $y<1$. We have

$$\zeta(s)\neq 0,\ \zeta(s)\ll \log^5|t| \quad \left(\sigma\geqslant 1-2c_1\frac{\log_2|t|}{\log|t|},\ |t|\geqslant t_0\right) \tag{3.28}$$

(Titchmarsh (1951), Theorem 5.17), and we deduce that

$$|\log\zeta(s)|\ll \log_2|t|,\quad \frac{\zeta'(s)}{\zeta(s)}\ll\frac{\log^2|t|}{\log_2|t|}, \tag{3.29}$$

for $\sigma\geqslant 1-c_1(\log_2|t|)/\log|t|$ $(|t|\geqslant t_1)$. To obtain the upper bound for $|\log\zeta(s)|$ we apply the Borel–Carathéodory inequality (Titchmarsh (1939), §5.5) with concentric circles centred at $1+it$, radii $r_1=1-\sigma, r_2=3r_1/2$, using the upper bounds $\operatorname{Re}\log\zeta(s)=\log|\zeta(s)|\leqslant\log\log|t|+O(1)$ on the large circle, $|\log\zeta(1+it)|\leqslant\log\log|t|+O(1)$ at the centre. To estimate the logarithmic derivative we apply Cauchy's formula

$$\frac{\zeta'(s)}{\zeta(s)}=\frac{1}{2\pi i}\int_D\frac{\log\zeta(z)\,dz}{(z-s)^2},$$

D is the circle $|z-s|=\frac14 c_1(\log_2|t|)/\log|t|$.

We move the path of integration in Perron's formula to $\sigma=1-c_2\log_2 t'/\log t'$, $t'=\max(|t|,t_1)$, where $c_2\leqslant c_1$, $c_2\log_2 t_1/\log t_1<\frac12$, joined up to $c\pm iT$ by horizontal segments, and with three lacets around the algebraic singularities of R. Note that $|\theta|\leqslant\exp\sqrt{(\log x)}<T/2=\frac12\exp\{4(\log x)/\log_2 x\}$ when x is large. Except on the lacets we have $R(s)\ll(\log T)^{20}$ and the contribution of this part of the contour is

$$\ll_\delta xe^{-\delta(\log\log x)^2}\log^{21}x\quad(\delta<\tfrac14 c_2),$$

which is negligible. On the lacet around $s=1$ we write $s^{-1}R(s)=:S(s)/(s-1)^{2y}$, where

$$S(s)=\frac1s H^2(s,y)\{(s-1)\zeta(s)\}^{2y}\frac{\zeta(s+i\theta,y)\zeta(s-i\theta,y)}{\zeta(2s,y^2)}$$

and we approximate to $S(s)$ with $S(1)$. We need an upper bound for $|S'(s)|$ and, differentiating S logarithmically, we find from (3.28), (3.29) that

$$S'(s)\ll\{|\theta|^{-1-2y}+\log^{10y+2}(3+|\theta|)\}.$$

An integration from s to 1 yields

$$S(s)=S(1)+O(|s-1|(|\theta|^{-1-2y}+\log^{10y+2}(3+|\theta|)))$$

and integrating the error term along the cut gives

$$\ll_y\{|\theta|^{-1-2y}+\log^{10y+2}(3+|\theta|)\}x\log^{2y-2}x.$$

For the main term, we extend the lacet to a loop from $-\infty$ around $s=1$.

The error at this point is negligible, and we obtain

$$\frac{S(1)x}{2\pi i}\int_{(-\infty)} x^{s-1}(s-1)^{-2y}\,ds = \frac{S(1)}{\Gamma(2y)}x(\log x)^{2y-1}$$

by Hankel's formula. (Whittaker & Watson (1927), §12.22). Since $H(1,y)=H(y)$, this is the main term in (3.21). The second term there arises from a similar calculation on the lacets around $1\pm i\theta$, the error involved being then

$$\ll x\{|\theta|^{-1-3y}+\log^{15y+2}(3+|\theta|)\}\log^{y-2}x$$

which does not exceed the error term exhibited in (3.21).

The left-hand sides of (3.21) and (3.22) are related via (3.18), but we cannot deduce (3.22) directly from (3.21). Rather, we proceed as above, replacing $R(s)$ in (3.27) by

$$R_1(s):=\frac{\zeta(s,y)\zeta(s+i\theta,y)\zeta(s-i\theta,y)}{\zeta(2s,y^2)}.$$

Notice that the three singularities now carry equal weight.

Notes on Chapter 3

We may write $\tau(n,\theta)$, rather clumsily, as $\sigma_{i\theta}(n)$, employing the familiar notation $\sigma_a(n)=\sum d^a$. The first genuine application of $\sigma_a(n)$ with complex a seems to be the proof that $\zeta(1+i\theta)\neq 0$ of Ingham (1930) (cf. Titchmarsh (1951), §3.4), which depends on Ramanujan's formula

$$\sum_{n=1}^{\infty}\frac{\sigma_a(n)\sigma_b(n)}{n^s}=\frac{\zeta(s)\zeta(s-a)\zeta(s-b)\zeta(s-a-b)}{\zeta(2s-a-b)}$$

(with $a=i\theta, b=-i\theta$: this is (3.27) with $y=1$), and a theorem of Landau. The same method is useful for $L(1+i\theta,\chi)\neq 0,\chi$ real.

Envisaged as a means of coping with divisors, $\tau(n,\theta)$ was formally defined in the modern notation by Hall (1975); but it had already appeared in Hall (1974), where it was used to show that, on a suitable sequence of integers n of asymptotic density 1, the set of real numbers $\{\log d: d|n\}$ is asymptotically uniformly distributed (mod 1). Indeed it follows from the upper bound for discrepancy of Erdős & Turán (1948) (cf. Kuipers & Niederreiter (1974)), that if we define

$$D(n):=\sup_{0\leqslant a<b\leqslant 1}\left|\sum_{d|n}\chi_{(a,b]}(\log d)-(b-a)\tau(n)\right|,$$

where $\chi_{(a,b]}$ is the characteristic function of the intervals $\equiv (a,b]$ (mod 1), then the *discrepancy* D satisfies $D(n)<\log^{\kappa+\varepsilon}n$ p.p., with $\kappa=\log(4/\pi)$. This result was obtained independently by Kátai (1976). It is an open problem whether it holds with $\kappa=0$.

The functions $(\log_2 d)^\alpha$ $(\alpha>1)$ (Hall (1978)), $(\log d)^\alpha$ $(\alpha>0)$, (Tenenbaum (1982)), λd(λ *irrational*) (Dupain, Hall & Tenenbaum (1982)) are all uniformly distributed (mod 1) in a similar sense, that is $D(n)=o(\tau(n))$ p.p. If we think of the divisors as having broadly exponential growth the results concerning $(\log_2 d)^\alpha$ and $(\log d)^\alpha$ above correspond very well to those for $(\log n)^\alpha, n^\alpha$ in $\mathbb{Z}^+$. In this scenario d^α grows very rapidly; it is shown to be uniformly distributed in Hall & Tenenbaum (1986): this paper gives a background and survey of the area.

Divisor density was defined by Hall (1978). If $\mathscr{A}$ is an integer sequence and $\tau(n,\mathscr{A})$ denotes the number of divisors d of n such that $d\in\mathscr{A}$, we say $\mathscr{A}$ has divisor density z, and write $\mathbf{D}\mathscr{A}=z$, if $\tau(n,\mathscr{A})\sim z\tau(n)$ p.p. For example, since $(\log d)^\alpha$ is uniformly distributed (mod 1), the sequence

$$\mathscr{A}(z):=\{d:(\log d)^\alpha\leqslant z \pmod 1\}\quad(0\leqslant z\leqslant 1)$$

has $\mathbf{D}\mathscr{A}=z$: notice that if $z>0$ we may also infer that the *set of multiples* $\mathscr{B}(\mathscr{A})$ has $\mathbf{d}\mathscr{B}=1$.

An intriguing conjecture of Erdős, related to that studied in Chapter 5, was that there exists $\lambda > 1$ such that almost all numbers have a divisor in at least one of the intervals $(\exp k^\lambda, 2\exp k^\lambda]$ $(k = 1, 2, 3, \ldots)$, and this was proved by Hall & Tenenbaum (1986), for all $\lambda < 1.314\,57\ldots$, using divisor density. The method is to set $\alpha = 1/\lambda$ in $\mathscr{A}(z)$ above, with $z \to 0$ essentially like $(\log n)^{\alpha - 1}$. Since $z\tau(n)$ is then relatively small (we only need it to be $\geqslant 1$) a good estimate for the discrepancy is required. An open problem is to determine how large λ may be in Erdős' conjecture.

Exercises on Chapter 3

30. Use the method of low moments and Theorem 01 to show that if $\xi(n)\to\infty$ then

$$(\log n)^{\log 2}\xi(n)^{-\sqrt{\log\log n}} < \tau(n) < (\log n)^{\log 2}\xi(n)^{\sqrt{\log\log n}} \quad \text{p.p.}$$

31. Let θ be fixed. Prove that infinitely often $|\tau(n,\theta)| > 2^{X_n}$ where

$$X_n = \frac{\log n}{\log_2 n} - \frac{c\log n}{(\log_2 n)^2}\log_3 n$$

and c is a suitable constant.

32. Let F be as in (3.8). Prove that for each $\delta>0$ the equation $\lambda F'(\lambda) - F(\lambda)+1=\delta$ has a unique positive root $\lambda=\lambda(\delta)$. Prove that if θ is fixed, $\theta\neq 0$, then, except for $\ll x/(\log x)^\delta$ exceptional integers $n\leqslant x$, we have $|\tau(n,\theta)|<(\log n)^D$, where $D=F'(\lambda(\delta))$.

33. Prove Theorem 31.

34. Prove that, for real t, and $\lambda>0$,

$$|1+e^{it}|^\lambda = F(\lambda) + \sum_{\substack{m=-\infty\\ m\neq 0}}^{\infty} a_m(\lambda)e^{imt}$$

where $a_m(\lambda)\ll |m|^{-1-\mu}$, $\mu=\min(1,\lambda)$. Prove that

$$A(\lambda,\theta) = AF(\lambda) - (2^\lambda - F(\lambda))\log^+\frac{1}{|\theta|} + \sum_{\substack{m=-\infty\\ m\neq 0}}^{\infty} a_m(\lambda)Z(m\theta)$$

where

$$A = \lim_{x\to 0}\left\{\sum_{p\leqslant x}\frac{1}{p} - \log_2 x\right\},\quad Z(\theta)=\sum_p p^{i\theta-1},\ \theta\neq 0.$$

35. Use the first part of Exercise 34 to show that, if $(a,b)=1$, $\max(a,b)>1$, then

$$\frac{1}{2\pi}\int_0^{2\pi} |(1+e^{iav})(1+e^{ibv})|\,dv \leqslant \Lambda < 2$$

where Λ is an absolute constant. Use Lemma 30.1 and Theorem 01 to show that if θ/φ is rational and not zero or ± 1 then

$$\sum_{n\leqslant x} |\tau(n,\theta)\tau(n,\varphi)| \ll x\log^{\Lambda-1} x.$$

36. Prove that as $\max(a,b)\to\infty$ the value of the integral in Exercise 35 converges to $16/\pi^2$. Find out how the $\ll$ in Exercise 35 depends on θ and φ when θ/φ is rational. Hence prove that, if θ/φ is irrational, then

$$\sum_{n\leqslant x} |\tau(n,\theta)\tau(n,\varphi)| \ll_\varepsilon x(\log x)^{16/\pi^2-1+\varepsilon}.$$

4
Measures of propinquity

4.1 Introduction

We investigate to what extent the divisors of a large 'random' integer n lie relatively close together. We define four functions, $T(n, \alpha)$, $U(n, \alpha)$, $\tau^+(n)$ and $\Delta(n)$, each of which measures in some sense the propinquity of the divisors, and we discuss relations between these functions. We are interested in p.p. upper and lower bounds where these are available, and suitably weighted average orders.

4.2 T and U – preliminary matters

For $\alpha \leqslant 1$ we set

$$T(n, \alpha) = \text{card}\,\{d, d' : d|n, d'|n, |\log(d/d')| \leqslant \log^\alpha n\}$$
$$= \sum_{\substack{\delta\delta'|n \\ (\delta,\delta')=1 \\ |\log(\delta/\delta')| \leqslant \log^\alpha n}} \tau(n/\delta\delta') \quad (n \geqslant 2), \tag{4.1}$$

$$U(n, \alpha) = \sum_{\substack{\delta\delta'|n \\ (\delta,\delta')=1 \\ |\log(\delta/\delta')| \leqslant \log^\alpha n}} 1 \quad (n \geqslant 2), \tag{4.2}$$

and $T(1, \alpha) = U(1, \alpha) = 1$ (for every α).

Notice that in the first line of (4.1) we have only to add the extra condition $(d, d') = 1$ to obtain $U(n, \alpha)$ instead of $T(n, \alpha)$. However the two functions behave quite differently. In particular $T(n, \alpha)$ counts all the pairs $d = d'$ which $U(n, \alpha)$ does not. By setting $t = (d, d')$ in (4.1) we see that, for $\alpha \geqslant 0$, we have

$$T(n, \alpha) \geqslant \sum_{t|n} U(n/t, \alpha) \tag{4.3}$$

with equality when $\alpha = 0$.

We have (cf. (3.15))

$$T(n, 1) = \tau(n)^2, \quad U(n, 1) = U(n) := \prod_{p^\nu \| n} (2\nu + 1); \tag{4.4}$$

of course we are more interested in the case $\alpha < 1$. We begin with an elementary result about $T(n, 0)$.

Theorem 40. *Let $m \in \mathbb{Z}^+$ and n be any multiple of m. Then*

$$1 \leqslant \frac{T(m,0)}{\tau(m)} \leqslant \frac{T(n,0)}{\tau(n)}. \tag{4.5}$$

Proof. The left-hand inequality is clear. For that on the right, observe first that, for any integers $a \geqslant 1$, $D \geqslant 1$, $d|D$,

$$\tau(aD)\tau(D)^{-1} \leqslant \tau(ad)\tau(d)^{-1}. \tag{4.6}$$

This follows immediately on writing the left-hand side as

$$\prod_{p^\nu \| a} \left(1 + \frac{\nu}{1 + v_p(D)}\right)$$

where $v_p(D)$ is the p-adic valuation of D. Now, from (4.1),

$$\frac{T(m,0)}{\tau(m)} = \frac{1}{\tau(m)} \sum_{\delta\delta'|m}{}^{*} \tau(m/\delta\delta'),$$

where the star indicates that the sum is restricted to those pairs δ, δ' for which $(\delta, \delta') = 1$, $|\log(\delta/\delta')| \leqslant 1$. We apply (4.6) with $a = n/m$, $D = m$, $d = m/\delta\delta'$ which yields $\tau(m/\delta\delta')/\tau(m) \leqslant \tau(n/\delta\delta')/\tau(n)$. Therefore

$$\frac{T(m,0)}{\tau(m)} \leqslant \frac{1}{\tau(n)} \sum_{\delta\delta'|m}{}^{*} \tau(n/\delta\delta') \leqslant \frac{T(n,0)}{\tau(n)}$$

as required.

Theorem 40 has an immediate consequence. For $X \geqslant 1$, put

$$\mathscr{M}(X) := \{n \in \mathbb{Z}^+ : T(n,0) > X\tau(n)\}.$$

Then, for each X, the logarithmic density $\boldsymbol{\delta}\mathscr{M}(X)$ *exists.*

To see this, notice that $\mathscr{M}(X)$ has the primitive sub-sequence $\mathscr{P}(X)$ comprising those $n \in \mathscr{M}$ such that $T(d,0)/\tau(d) \leqslant X$ for every proper divisor d of n. By the theorem, $\mathscr{M}(X)$ is just the set of multiples of $\mathscr{P}(X)$. An important theorem of Davenport & Erdős (1937) (cf. Halberstam & Roth (1966), Chapter V) states that any set of multiples possesses logarithmic density (equal to its lower asymptotic density); and this implies our result. More is true: this will emerge later.

The function $T(n,0)$ is important in the study of Erdős' τ^+-function and Hooley's Δ-function, both to be defined later. However, the contribution to $T(n,\alpha)$ from the admissible coprime pairs δ, δ' for which $\tau(n/\delta\delta')$ is large (e.g. $\delta = \delta' = 1$) tends to be rather over-emphasized when α is small, and for this reason when considering $T(n,\alpha)$ we restrict our attention to the range $\alpha \geqslant 0$. The definition of U avoids this phenomenon and some very interesting questions arise when $\alpha < 0$.

4.3 T and U – average orders

The average orders of divisor functions of this sort tend to be dominated by the contribution from integers with abnormally many, or abnormally distributed, prime factors. We can reduce this effect by introducing a weight $y^{\Omega(n)}$ with $y < 1$. This device can be particularly helpful when we seek p.p. upper bounds but it should be emphasized that, with functions like T and U which depend on the distribution of prime factors as well as their number, we can only go so far in this direction with such a simple weight.

Theorem 41. *Let* $0 < \alpha < 1$, $0 < y < \frac{1}{2}$. *Then*

$$\sum_{n \leqslant x} T(n, \alpha) y^{\Omega(n)} = x(\log x)^{(2\alpha+2)y-1}(T_0(y) + O((\log x)^{-\kappa})), \tag{4.7}$$

where

$$\kappa = \kappa(\alpha, y) := \min\{(1-\alpha)(1-2y), (1+2\alpha)y, 2\alpha/(2-y)\},$$

$$T_0(y) := \frac{\Gamma(1-2y)H^4(y)}{\Gamma(2y)\Gamma(1-y)\Gamma(1+y)\zeta(2, y^2)},$$

and $\zeta(s, y)$, $H(y)$ *are as in* (3.23), (3.24). *Also*

$$\sum_{n \leqslant x} U(n, \alpha) y^{\Omega(n)} = x(\log x)^{(2\alpha+1)y-1}(U_0(y) + O((\log x)^{-\kappa'})), \tag{4.8}$$

where

$$\kappa' = \kappa'(\alpha, y) := \min\{(1-\alpha)(1-2y), 2\alpha y\},$$

$$U_0(y) := \frac{\Gamma(1-2y)H^3(y)}{\Gamma(y)\Gamma(1-y)\Gamma(1+y)\zeta(2, y^2)}.$$

The reader will observe that, if (4.3) had been an equality, the ratio between the main terms in (4.7) and (4.8) would have been $(H(y)\Gamma(2\alpha y + y)/\Gamma(2\alpha y + 2y))\log^y x$ whereas in fact $T_0(y)/U_0(y) = H(y)\Gamma(y)/\Gamma(2y)$, revealing that, as we might expect, the inequality in (4.3) becomes more pronounced as α increases, but not enough to upset the orders of magnitude of the main terms.

Theorem 42. *Let* $0 \leqslant \alpha < 1$. *Then*

$$\sum_{n \leqslant x} T(n, \alpha)/2^{\Omega(n)} = \frac{2H^4(\frac{1}{2})}{\pi\zeta(2, \frac{1}{4})}(1-\alpha)x\log^\alpha x \cdot \log_2 x + O\{x\log^\alpha x + x(\log_3 x)^2\} \tag{4.9}$$

and

$$\sum_{n \leqslant x} U(n, \alpha)/2^{\Omega(n)} = \frac{2H^3(\frac{1}{2})}{\pi^{3/2}\zeta(2, \frac{1}{4})}(1-\alpha)x\log^{\alpha-1/2} x \cdot \log_2 x + O\{x\log^{\alpha-1/2} x + x(\log_3 x)^2/\sqrt{\log x}\}. \tag{4.10}$$

In both these theorems we have excluded the case $\alpha=1$: $T(n,1)$ and $U(n,1)$ are multiplicative functions (cf. (4.4)) for which it is a straightforward matter to write down the corresponding asymptotic formulae, with very good error terms and main terms which are not obtained by substituting $\alpha=1$ in (4.7)–(4.10). This discontinuity in the main terms explains the deterioration in the quality of the error terms above when $\alpha\to 1$.

We may choose y optimally as a function of α to obtain p.p. upper bounds.

Corollary. *Let* $\alpha\in(0,1)$, $\alpha\in(\frac{1}{2},1)$ *respectively. Then*

$$\sum_{n\leqslant x}\frac{T(n,\alpha)}{(2\alpha+2)^{\Omega(n)}}=T_1(\alpha)x+O(x(\log x)^{-\alpha(1-\alpha)/(1+\alpha)}), \tag{4.11}$$

$$\sum_{n\leqslant x}\frac{U(n,\alpha)}{(2\alpha+1)^{\Omega(n)}}=U_1(\alpha)x+O(x(\log x)^{-(2\alpha-1)(1-\alpha)/(1+2\alpha)}), \tag{4.12}$$

where $T_1(\alpha)=T_0(1/(2\alpha+2))$, $U_1(\alpha)=U_0(1/(2\alpha+1))$.

Furthermore, if $\xi(n)\to\infty$ *as* $n\to\infty$ *we have, for* $0\leqslant\alpha\leqslant 1$,

$$T(n,\alpha)<(\log n)^{\log(2\alpha+2)+\xi(n)/\sqrt{\log\log n}}\quad \text{p.p.}, \tag{4.13}$$

$$U(n,\alpha)<(\log n)^{l(\alpha)+\xi(n)/\sqrt{\log\log n}}\quad \text{p.p.}, \tag{4.14}$$

where

$$l(\alpha)=\begin{cases}\alpha-\frac{1}{2}+\log 2 & (0\leqslant\alpha\leqslant\frac{1}{2}),\\ \log(2\alpha+1) & (\frac{1}{2}<\alpha\leqslant 1).\end{cases}$$

We obtain (4.11) from (4.7) and (4.12) from (4.8); notice that we require $\alpha>\frac{1}{2}$ in (4.12) to make $y=(2\alpha+1)^{-1}<\frac{1}{2}$. To obtain (4.13), we insert the p.p. upper bound for $\Omega(n)$ from the Hardy–Ramanujan Theorem into (4.9) ($\alpha=0$) and (4.11) ($0<\alpha<1$). Similarly (4.14) depends on (4.10) ($0\leqslant\alpha\leqslant\frac{1}{2}$) and (4.12) ($\frac{1}{2}<\alpha<1$). When $\alpha=1$ we employ (4.4).

The exponent of $\log n$ in (4.13) is sharp when $\alpha=0$ or 1, and that in (4.14) when $\alpha=1$. We determine the normal orders of $\log T(n,\alpha)$, $\log U(n,\alpha)$ later (cf. Theorems 44, 45 and 53), and except in the cases mentioned we see that the exponents in (4.13), (4.14) are not sharp, although the true values are not exceeded by as much as 0.06, 0.1 respectively for any α.

There are asymptotic formulae corresponding to those given in Theorem 41, valid for $y\in(\frac{1}{2},2)$ (for larger y we should have to replace Ω by ω), in which the exponents of $\log x$ become $4y-2+\alpha$ in (4.7) and $3y-2+\alpha$ in (4.8). $T_0(y)$ and $U_0(y)$ alter as well. A different method is needed when $y>\frac{1}{2}$, and we do not treat this case in our book (see the Notes). However we quote a particular formula of this sort without proof – it should be compared with (4.12).

Theorem 43. *Let* $-1 \leqslant \alpha < \frac{1}{2}$. *Then*

$$\sum_{n \leqslant x} U(n,\alpha)\left(\frac{2-\alpha}{3}\right)^{\Omega(n)} \sim U_1(\alpha)x,$$

where now

$$U_1(\alpha) = \frac{4^{1-\beta}\Gamma(2\beta-1)H^3(\beta)}{\Gamma^2(\beta)\Gamma(3\beta-1)\zeta(2,\beta^2)} \quad (\beta=(2-\alpha)/3).$$

In particular the average value of $U(n,-1)$ is $6/\pi^2$.

The proofs of Theorems 41 and 42 depend on Theorem 32, the main idea being to approximate $T(n,\alpha)$ or $U(n,\alpha)$ by Fourier integrals involving the functions $|\tau(n,\theta)|^2$ or $\rho(n,\theta)$ respectively – see (4.18). We need a result which copes efficiently with the oscillating term involving $x^{i\theta}$ in (3.21) and (3.22). This is the following.

Lemma 41.1. *Let* $0 < \operatorname{Re} z < 1$. *Then, for real* $w \neq 0$,

$$\int_0^\infty \frac{\sin t}{t^{1+z}} e^{itw}\, dt \ll_z |w|^{\operatorname{Re} z - 1}. \tag{4.15}$$

In addition, for $w > 0$ *we have*

$$\int_{1/w}^\infty \frac{\sin t}{t^2} e^{itw}\, dt = O(1). \tag{4.16}$$

Proof of (4.15). We may assume $|w| \geqslant 2$. The integral equals

$$\begin{aligned} &\frac{1}{\Gamma(z)}\int_0^\infty \frac{\sin t}{t} e^{itw} \int_0^\infty e^{-tv} v^{z-1}\, dv\, dt \\ &= \frac{1}{\Gamma(z)}\int_0^\infty v^{z-1} \arctan\left(\frac{1}{v-iw}\right) dv \\ &\ll_z \int_0^\infty \frac{v^{\operatorname{Re} z - 1}}{|v-iw|}\, dv \ll_z |w|^{\operatorname{Re} z - 1}. \end{aligned}$$

Proof of (4.16). We may assume $w \geqslant 2$. The integral equals

$$\frac{1}{2i}\{g(w+1) - g(w-1) + O(1)\}$$

where

$$g(w) := \int_{1/w}^\infty e^{itw} \frac{dt}{t^2} = wg(1).$$

The result follows.

Proofs of Theorems 41 and 42. We shall concentrate on $T(n,\alpha)$, $U(n,\alpha)$ being similar except for one point which requires explanation. We need formula (3.21) for T and (3.22) for U: in each formula there are really two main terms, the second involving $\cos(\theta\log x)$ and $\sin(\theta\log x)$ and so rapidly oscillating as θ varies. It is treated as an error term, and this needs some care in dealing with U because the two main terms in (3.22), unlike (3.21), have the same order of magnitude. The same analysis might just as well be applied to T: hence the slightly better error term in this case.

For $h, \eta > 0$ put $f(u; h, \eta) = 1$ if $|u| \leqslant h$, else $f(u; h, \eta) = (1 - \eta^{-1}(|u| - h))^+$. The graph is a trapezium supported on $[-h-\eta,\ h+\eta]$. The Fourier transform of f is

$$\hat{f}(\theta; h, \eta) = \frac{2}{\eta\theta^2}(\cos h\theta - \cos(h+\eta)\theta) \ll \min\{h+\eta, 1/|\theta|, 1/\eta|\theta|^2\}. \quad (4.17)$$

We define

$$W(n; h, \eta) = \sum_{d|n}\sum_{d'|n} f(\log(d/d'); h, \eta) = \frac{1}{2\pi}\int_{-\infty}^{\infty} \hat{f}(\theta; h, \eta)|\tau(n, \theta)|^2\, d\theta; \quad (4.18)$$

if $h \doteqdot \log^\alpha n$ and η is small this approximates $T(n, \alpha)$. Notice that when we come to $U(n, \alpha)$ the extra summation condition $(d, d') = 1$ is needed in (4.18) and $|\tau(n, \theta)|^2$ has to be replaced by $\rho(n, \theta)$.

Put $h_1 = \log^\alpha(x/\log x)$, $h_2 = \log^\alpha x$. We have

$$\begin{aligned} T(n, \alpha) &\leqslant W(n; h_2, \eta) \qquad (n \leqslant x) \\ &\geqslant W(n; h_1 - \eta, \eta) \quad (x/\log x < n \leqslant x), \end{aligned} \quad (4.19)$$

and we set $\eta = 1/\log_2 x$ if $\alpha = 0$. For $\alpha > 0$ it will be chosen later, from the range $(1/\log x, 1)$. In what follows, h will be either $h_1 - \eta$ or h_2. We deduce from (4.18) that

$$\sum_{n \leqslant x} W(n; h, \eta) y^{\Omega(n)} = \frac{1}{2\pi}\int_I \hat{f}(\theta; h, \eta) \sum_{n \leqslant x} |\tau(n, \theta)|^2 y^{\Omega(n)}\, d\theta + O(hx\log^{4y-2} x),$$

where $I := \{\theta: 1/\log x < |\theta| < \exp\sqrt{\log x}\}$. The error term follows from (4.17) and the upper bound

$$\sum_{n \leqslant x} |\tau(n, \theta)|^2 y^{\Omega(n)} \leqslant \sum_{n \leqslant x} \tau(n)^2 y^{\Omega(n)} \ll x(\log x)^{4y-1}$$

(which is a consequence of Theorem 01). Notice that it has the order of magnitude of the main term which one obtains when $y > \frac{1}{2}$, restricting this method. We insert (3.21) so that the right-hand side becomes

$$\frac{1}{2\pi}\int_I \hat{f}(\theta; h, \eta)(\text{Main Terms})\, d\theta + O(hx\log^{4y-2} x + \log^{13y+3}(2/\eta)\cdot x\log^{2y-2} x). \quad (4.20)$$

Since $y > 0$ and $h \gg 1$ the right-hand error term here is negligible. We begin with the second (oscillating) main term.

We have to estimate

$$x \log^{y-1} x \int_I \frac{\zeta^2(1+i\theta, y)\zeta(1+2i\theta, y)}{(1+i\theta)\zeta(2+2i\theta, y^2)} x^{i\theta} \hat{f}(\theta; h, \eta)\, d\theta$$

and we notice first that, since $\zeta(1+i\theta) \ll \log|\theta|$ for large $|\theta|$, the contribution from the range $|\theta| > \frac{1}{5}$ is $\ll x \log^{y-1} x$. Now consider the range $1/\log x < |\theta| \leqslant \frac{1}{5}$: we may suppose $\theta > 0$. We have (say)

$$\frac{\zeta^2(1+i\theta, y)\zeta(1+2i\theta, y)}{(1+i\theta)\zeta(2+2i\theta, y^2)} = \theta^{-3y}(c_0 + c_1\theta + c_2\theta^2 + \cdots)$$

where the infinite series has radius of convergence $\frac{1}{4}$: recall that $\zeta(s, y) = H(s, y)\zeta(s)^y$ where $H(s, y)$ is analytic for $\mathrm{Re}\, s > \frac{1}{2}$ if $y \leqslant 1$. If $y > \frac{1}{3}$, we apply (4.17) to obtain

$$\ll x \log^{y-1} x \int_{1/\log x}^{1/5} h\theta^{-3y}\, d\theta \ll_y hx \log^{4y-2} x.$$

Let $y \leqslant \frac{1}{3}$. The terms from $c_1\theta$ onward contribute

$$\ll (\log 2h)^{\delta(3y,1)} x \log^{y-1} x \tag{4.21}$$

and we are left with

$$c_0 x \log^{y-1} x \cdot \int_{1/\log x}^{1/5} \frac{4}{\eta\theta^2} \sin\left(h + \frac{\eta}{2}\right)\theta \cdot \sin\frac{\eta\theta}{2} \cdot \theta^{-3y} x^{i\theta}\, d\theta.$$

We approximate $\sin(\eta\theta/2)$ by $\eta\theta/2$: the error is absorbed by (4.21). Similarly the upper limit of integration may be changed to ∞. If $y = \frac{1}{3}$, we substitute $\theta = t/(h + \eta/2)$, put $w = (\log x)/(h + \eta/2)$ and apply Lemma 41.1 (4.16). We obtain $\ll hx \log^{y-1} x$, which absorbs (4.21), and may be written as $\ll hx \log^{4y-2} x$. If $y < \frac{1}{3}$, we change the lower limit of integration to 0, introducing an error $\ll hx \log^{4y-2} x$, and then apply Lemma 41.1 (4.15) with t and w as above, obtaining $\ll hx \log^{4y-2} x$ again. Putting these results together we deduce that the total contribution to (4.20) from the second main term is

$$\ll hx \log^{4y-2} x + x \log^{y-1} x \quad (0 < y \leqslant \tfrac{1}{2}).$$

Moreover this absorbs all the previous error terms.

The first main term in (4.20) is

$$\frac{H^2(y) x \log^{2y-1} x}{2\pi\Gamma(2y)\zeta(2, y^2)} \int_I |\zeta(1+i\theta, y)|^2 \hat{f}(\theta; h, \eta)\, d\theta \tag{4.23}$$

and we write

$$|\zeta(1+i\theta, y)|^2 =: \frac{H^2(y)}{|\theta|^{2y}} + Z(\theta, y). \tag{4.24}$$

We have

$$Z(\theta, y) = |\theta|^{-2y}(b_2\theta^2 + b_4\theta^4 + \cdots) \quad (|\theta| < \tfrac{1}{2}),$$

and we employ this for $|\theta| \leqslant \frac{1}{3}$; from (4.24) we have $Z(\theta, y) \ll \log^{2y}(6|\theta|)$, $Z'(\theta, y) \ll \log^{2y+1}(6|\theta|)$ when $|\theta| > \frac{1}{3}$.

We begin by estimating the integral involving $Z(\theta, y)$. From (4.17) and the above,

$$\int_I Z(\theta, y)\hat{f}(\theta; h, \eta)\, d\theta \ll \log^{2y+1}(2/\eta). \tag{4.25}$$

When $\alpha = 0$ (in Theorem 42) we put $\eta = 1/\log\log x$ so that this is $\ll (\log_3 x)^2$. When $\alpha > 0$, we can improve on (4.25). Put $\lambda = (\eta \log^\alpha x)^{-1/2}$ and assume that $\lambda \leqslant \frac{1}{3}$. Then

$$\int_{1/\log x}^{\lambda} Z(\theta, y)\hat{f}(\theta; h, \eta)\, d\theta \ll \int_0^{\lambda} \theta^{1-2y}\, d\theta \ll \lambda^{2-2y}$$

and an integration by parts yields

$$\int_{\lambda}^{\exp\sqrt{\log x}} Z(\theta, y)\frac{\cos h\theta}{\eta\theta^2}\, d\theta \ll \eta^{-1}h^{-1}\lambda^{-2y}$$

so that we obtain

$$\int_I Z(\theta, y)\hat{f}(\theta; h, \eta)\, d\theta \ll (\eta \log^\alpha x)^{y-1}. \tag{4.26}$$

Next, we consider the leading term in (4.24). If $y < \frac{1}{2}$, we extend the range of integration in (4.23) to $(-\infty, \infty)$ (for this term only), which introduces an error $\ll hx\log^{4y-2} x$, absorbed by (4.22). We then obtain the integral

$$4H^2(y)\eta^{-1}\int_0^{\infty} \{\cos h\theta - \cos(h+\eta)\theta\}\theta^{-2-2y}\, d\theta$$

$$= 4CH^2(y)\frac{(h+\eta)^{2y+1} - h^{2y+1}}{(2y+1)\eta} = 4CH^2(y)h^{2y} + O(\eta h^{2y-1})$$

where

$$C = (2y+1)\int_0^{\infty} \frac{1-\cos t}{t^{2+2y}}\, dt = \frac{\pi\Gamma(1-2y)}{2\Gamma(1-y)\Gamma(1+y)}.$$

Since $y < \frac{1}{2}$ we have only the case $\alpha > 0$ to consider. We put $\eta = (\log x)^{-\alpha y/(2-y)}$ and combine the above with (4.26) to deduce that the first main term (4.23) equals

$$T_0(y)h^{2y}x\log^{2y-1} x + O(hx\log^{4y-2} x + x\log^{2y-1-Y} x) \tag{4.27}$$

where $Y = 2\alpha(1-y)^2/(2-y)$. We put this together with (4.20) and (4.22) to obtain, for $y<\frac{1}{2}$,

$$\sum_{n\leqslant x} W(n;h,\eta)y^{\Omega(n)} = T_0(y)h^{2y}x\log^{2y-1}x + O(x\log^{4y-2+\alpha}x$$
$$+ x\log^{y-1}x + x\log^{2y-1-Y}x). \tag{4.28}$$

Now let $y=\frac{1}{2}$. Since we already have an error term $O(hx)$ in (4.20), (4.22) an error $O(h)$ in the evaluation of the integral in (4.23) would be acceptable: we cannot quite achieve this when $\alpha=0$ but when $\alpha>0$ it is easy to do better (which maybe points to a reappraisal of our method in this case).

We insert the leading term from (4.24) into the integral in (4.23) to obtain

$$4H^2(\tfrac{1}{2})\eta^{-1}\int_{1/\log x}^{1/\eta}\{\cos h\theta - \cos(h+\eta)\theta\}\frac{d\theta}{\theta^3} + O(h^{-1}\eta^2)$$
$$= 4H^2(\tfrac{1}{2})\int_{1/\log x}^{1/\eta}\sin\left(h+\frac{\eta}{2}\right)\theta\cdot\frac{d\theta}{\theta^2} + O(h^{-1}\eta^2),$$

replacing $\sin(\eta\theta/2)$ by $\eta\theta/2$ and estimating the error by an integration by parts. In the present treatment, we shall not be able to maintain the error term on the right. Put

$$C_1 = \int_0^1 \frac{\sin t - t}{t^2}\,dt + \int_1^\infty \frac{\sin t}{t^2}\,dt = 1-\gamma$$

(where γ denotes Euler's constant), $w=\log x/(h+\eta/2)$, and substitute $\theta = t/(h+\eta/2)$, to obtain

$$4H^2(\tfrac{1}{2})\left(h+\frac{\eta}{2}\right)(\log w + C_1 + O(w^{-2})) + O(h^{-1}\eta^2). \tag{4.29}$$

Recall that $h=h_1-\eta$ or h_2. In either case

$$\log w = (1-\alpha)\log_2 x + O(h^{-1}\eta + \log_2 x/\log x),$$

whence (4.29) becomes

$$4H^2(\tfrac{1}{2})h\{(1-\alpha)\log_2 x + 1 - \gamma\}$$
$$+ O\{\eta\log_2 x + h^3\log^{-2}x + h\log_2 x/\log x\}.$$

We have to consider two cases according as $\alpha>0$ or not. In the first case, there is an error $(\eta\log^\alpha x)^{-1/2}$ from (4.26), and we choose

$$\eta = \log^{-\alpha/3}x\cdot(\log_2 x)^{-2/3}.$$

The first main term (4.23) becomes

$$\frac{2H^4(\frac{1}{2})}{\pi\zeta(2,\frac{1}{4})}hx\{(1-\alpha)\log_2 x + 1 - \gamma\}$$
$$+ O(x\log^{-\alpha/3}x(\log_2 x)^{1/3} + x\log^{3\alpha-2}x)$$

(the error term $O(hx\log_2 x/\log x)$ being absorbed). Therefore

$$\sum_{n\leqslant x} W(n;h,\eta)2^{-\Omega(n)} = \frac{2H^4(\frac{1}{2})}{\pi\zeta(2,\frac{1}{4})} hx\{(1-\alpha)\log_2 x + O(1)\}. \tag{4.30}$$

In the second case, $\alpha = 0$, we put $\eta = 1/\log_2 x$. There is an error $\ll x(\log_3 x)^2$ from (4.25) and the $O(1)$ in (4.30) becomes $O((\log_3 x)^2)$. Now we put $h = h_2 = \log^\alpha x$ and recall (4.19): (4.28) and (4.30) yield (4.7) and (4.9) respectively with $\leqslant$. Put $h = h_1 - \eta$. If $y < \frac{1}{2}$,

$$\begin{aligned} h^{2y} &= h_1^{2y} + O(\eta h_1^{2y-1}) \\ &= \log^{2\alpha y} x + O(\log^{-Y} x + (\log x)^{2\alpha y - 1}\log_2 x) \end{aligned} \tag{4.31}$$

with Y as in (4.27), and (4.28) gives

$$\sum_{n\leqslant x} W(n;h_1-\eta,\eta)y^{\Omega(n)} = x\log^{(2\alpha+2)y-1}(T_0(y) + (\log x)^{-\kappa}),$$

the second error term in (4.31) being absorbed because $(2\alpha+2)y - 2 < 4y - 2 + \alpha$. For the same reason, that part of the sum on the left above over the range $n \leqslant x/\log x$ is negligible, and for $x/\log x < n \leqslant x$ we may apply (4.19) to deduce (4.7) with $\geqslant$. We obtain (4.9) with $\geqslant$ from (4.30) in a similar fashion.

We indicate briefly the changes needed to prove the corresponding results for $U(n,\alpha)$. The error involved in the formula parallel to (4.20) together with the contribution from the second main term is

$$\ll hx\log^{3y-2} x + x\log^{y-1} x$$

and comparing this with (4.22) we see that the term on the right is unchanged and is now relatively more serious. Indeed the first main term, parallel to (4.23), is

$$\frac{H(y)x\log^{y-1} x}{2\pi\Gamma(y)\zeta(2,y^2)} \int_I |\zeta(1+i\theta,y)|^2 \hat{f}(\theta;h,\eta)\,d\theta$$

and we can afford a simpler treatment of the integral on the right.

4.4 The normal orders of log T and log U

It may be as well to recall the definition of the normal order of an arithmetical function of Hardy & Ramanujan (1917): '*If then $g(n)$ is an arithmetical function of n, and $\varphi(n)$ an elementary increasing function, and if, for every positive ε, we have*

$$(1-\varepsilon)\varphi(n) < g(n) < (1+\varepsilon)\varphi(n)$$

for almost all values of n, we shall say that the normal order of $g(n)$ is $\varphi(n)$'.

Thus $\tau(n)$ does not have a normal order (cf. Exercise 42). However it has become traditional to say rather loosely that $\tau(n)$ has normal order $(\log n)^{\log 2}$ when we mean

$$\tau(n) = (\log n)^{\log 2 + o(1)} \quad \text{p.p.},$$

and we can talk about the 'normal order of $T(n, \alpha)$' in a similar cavalier fashion. The idea is useful, and this is what matters; but we stress from the outset that it is $\log T(n, \alpha)$ and $\log U(n, \alpha)$ that have normal orders (for suitable values of α).

Theorem 44. *Let* $\alpha \in [0, 1]$. *The function* $\log T(n, \alpha)$ *has normal order* $(1 + \alpha)\log 2 \cdot \log_2 n$, *more precisely if* $\xi(n) \to \infty$ *as* $n \to \infty$ *then*

$$\frac{\log T(n, \alpha)}{\log_2 n} = (1 + \alpha)\log 2 + O\left(\frac{\xi(n)}{\sqrt{\log_2 n}}\right) \quad \text{p.p.} \tag{4.32}$$

This should be compared with (4.13).

Proof. Let $m|n$ and $\log m \leqslant \log^\alpha n$. If $\delta|m$, $\delta'|m$, $t|n/m$ then δt, $\delta' t$ are a pair of divisors counted by $T(n, \alpha)$, and so

$$T(n, \alpha) \geqslant \tau(m)^2 \tau(n/m) \geqslant \tau(n)\tau(m) \geqslant 2^{\omega(n) + \omega(m)}.$$

Let m be the product of all the prime factors p of n such that $\log p \leqslant \log^\alpha n / 2\log_2 n$. Since $\Omega(n) \leqslant 2\log_2 n$ p.p. we have $\log m \leqslant \log^\alpha n$ p.p. and, by Theorem 010, both

$$\omega(n) \geqslant \log_2 n - \xi(n)\sqrt{\log_2 n} \quad \text{p.p.},$$

and

$$\omega(m) \geqslant \alpha \log_2 n - \xi(n)\sqrt{\log_2 n} \quad \text{p.p.},$$

from which (4.32), with $\geqslant$, follows.

Now put $w = \exp \log^\alpha x$ and for each n define

$$n(\alpha) = \prod_{\substack{p^\nu \| n \\ p \leqslant w}} p^\nu.$$

Let $d, d' | n$, $|\log(d/d')| \leqslant \log^\alpha n$, and set $\delta = (n(\alpha), d)$, $\delta' = (n(\alpha), d')$, $t = d/\delta$, $t' = d'/\delta'$. Then

$$\begin{aligned} |\log(t/t')| &\leqslant |\log(d/d')| + |\log(\delta/\delta')| \\ &\leqslant \log^\alpha n + \Omega(n(\alpha)) \log^\alpha x \leqslant h \end{aligned}$$

for almost all integers $n \leqslant x$, with $h := 2\log^\alpha x \cdot \log_2 x$. It follows that, with W as in (4.18), for these integers we have

$$T(n, \alpha) \leqslant \tau^2(n(\alpha))\, W(n/n(\alpha); h, 1). \tag{4.33}$$

We consider the sum

$$\sum_{n\leqslant x} \tau(n/n(\alpha))^{-1} W(n/n(\alpha); h, 1)$$

$$= \frac{1}{2\pi}\int_{-\infty}^{\infty} \hat{f}(\theta; h, 1) \sum_{n\leqslant x} \tau(n/n(\alpha))^{-1}\left|\tau\left(\frac{n}{n(\alpha)}, \theta\right)\right|^2 d\theta$$

$$\ll x\log^{\alpha-1} x\cdot \int_0^{\infty} |\hat{f}(\theta; h, 1)| \exp\left\{\frac{1}{2}\sum_{w<p\leqslant x} p^{-1}|\tau(p,\theta)|^2\right\} d\theta \quad (4.34)$$

by Theorem 01. We split this integral into four ranges. The first is $[0, 1/\log x)$: its contribution is

$$\ll x\log^{\alpha-1} x\cdot \int_0^{1/\log x} h\exp\{2(1-\alpha)\log_2 x\}\, d\theta \ll x\log_2 x,$$

using (4.17). The second range is $[\log^{-1} x, \log^{-\alpha} x)$, and we notice that in this range, $w < \exp(1/\theta) \leqslant x$. We have

$$\frac{1}{2}\sum_{w<p\leqslant x} p^{-1}|\tau(p,\theta)|^2 \leqslant 2\left(\log\frac{1}{\theta} - \log_2 w\right) + O(1) + \frac{1}{2}\sum_{\exp(1/\theta)<p\leqslant x} p^{-1}|\tau(p,\theta)|^2$$

and we apply (3.9) to the sum on the right to obtain altogether

$$\leqslant \log(1/\theta) + (1-2\alpha)\log_2 x + O(1).$$

Hence the contribution from the second range is

$$\ll x\log^{\alpha-1} x\cdot \int_{1/\log x}^{\log^{-\alpha} x} h\log^{1-2\alpha} x\cdot\frac{d\theta}{\theta} \ll x(\log_2 x)^2.$$

The third range is $[\log^{-\alpha} x, 1)$. Since $w \geqslant \exp(1/\theta)$ here, (3.9) gives

$$\frac{1}{2}\sum_{w<p\leqslant x} p^{-1}|\tau(p,\theta)|^2 = (1-\alpha)\log_2 x + O(1)$$

and we obtain

$$\ll x\log^{\alpha-1} x\cdot \int_{\log^{-\alpha} x}^{1} \theta^{-1}\log^{1-\alpha} x\cdot d\theta \ll x\log_2 x.$$

In the fourth range $[1, \infty)$ this θ^{-1} becomes θ^{-2} giving $\ll x$. The sum (4.34) is $\ll x(\log_2 x)^2$ and we deduce that

$$W(n/n(\alpha); h, 1) < \tau(n/n(\alpha))(\log_2 n)^3$$

for all but $o(x)$ integers $n \leqslant x$. Together with (4.33) this implies

$$T(n,\alpha) \leqslant 2^{\Omega(n)+\Omega(n(\alpha))}(\log_2 n)^3 \quad \text{p.p.}$$

from which (4.32), with $\leqslant$, readily follows. This completes the proof of Theorem 44.

Now we consider $U(n,\alpha)$. We shall prove in this and the next chapter that, for $0 \leqslant \alpha \leqslant 1$, $\log U(n,\alpha)$ has normal order $(\log 3 - 1 + \alpha)\log n$. Thus $U(n,0) > (\log n)^{\log 3 - 1 - \varepsilon}$ p.p., in particular for almost all n there exist divisors d, d' such that $d < d' < ed$. Apart from the constant, this is Erdős' conjecture referred to in the introduction. Because of this, the p.p. lower bound for $U(n,\alpha)$ belongs in Chapter 5 (cf. Theorem 53), and we confine our attention to the upper bound here.

Theorem 45. *Let* $0 \leqslant \alpha \leqslant 1$, *and* $B > 3^{\sqrt{2}}$. *Then*

$$U(n,\alpha) < (\log n)^{\log 3 - 1 + \alpha} B^{\sqrt{(\log_2 n \cdot \log_4 n)}} \quad \text{p.p.} \tag{4.35}$$

This should be compared with (4.14). Of course when $\alpha = 1$ a slightly stronger (and sharp) upper bound is available from the Hardy–Ramanujan Theorem: from (4.4) we have $3^{\omega(n)} \leqslant U(n,1) \leqslant 3^{\Omega(n)}$.

The exponent $\log 3 - 1 + \alpha$ is suggested by a simple heuristic argument. The numbers $\log(d/d')$ where $d, d'|n$, $(d,d') = 1$ are distinct and lie in the interval $[-\log n, \log n]$, $U(n,\alpha)$ of them in $[-\log^\alpha n, \log^\alpha n]$. We might expect the proportion of these numbers in the shorter interval to be about $\log^{\alpha-1} n$, that is

$$U(n,\alpha) \doteqdot U(n,1)\log^{\alpha-1} n,$$

which would lead to the correct result.

The normal order of $\log T(n,\alpha)$ is not $(\log 4 - 1 + \alpha)\log_2 n$ (unless $\alpha = 1$) – of course in this case the numbers $\log(d/d')$ are not distinct; but the difference in the behaviour of T and U lies deeper than this. Write $n = PQ$ where P is the product of all the prime factors $\leqslant \exp(\log^\alpha n)$. Then normally $T(n,\alpha)$ behaves like $\tau^2(P)\tau(Q)$, that is a typical pair of divisors counted by $T(n,\alpha)$ will have (more or less!) $(d,Q) = (d',Q)$. If such were true of $U(n,\alpha)$ it would behave like $U(P,1)$, and the normal order of $\log U(P,1)$ is easily seen to be $\alpha \log 3 \cdot \log_2 n$. To fly higher than this $U(n,\alpha)$ needs a substantial involvement from the 'large' prime factors.

Lemma 45.1. *Let* $0 < y < 2$, $Y \geqslant 0$, $\frac{2}{3} \leqslant u < v \leqslant x$. *Then*

$$\sum_{u<d\leqslant v} d^{-1} y^{\Omega(d)} \left(\log \frac{2x}{d}\right)^{-Y} \ll (\log 2u)^{y-1} \int_u^v \left(\log \frac{2x}{t}\right)^{-Y} \frac{dt}{t} + (\log 2u)^{y-2} \log^{-Y}(2x/v).$$

This follows from Theorem 04.

Proof of Theorem 45. The Fourier transform method is not applicable here (without serious modification) because $\rho(n,\theta)$ is not positive.

We put $z = \exp \log^{\alpha} x$ and define

$$V_1(n) = \sum \{1: dd'|n, d < d' \leqslant dz, d \leqslant x^{1/4}\},$$
$$V_2(n) = \sum \{1: dd'|n, d < d' \leqslant dz, d > x^{1/4}\}.$$

Plainly $U(n, \alpha) \leqslant 1 + 2(V_1(n) + V_2(n))$. We begin with an upper bound, valid for $y \leqslant 1$, for

$$S_2 := \sum_{n \leqslant x} V_2(n) y^{\Omega(n)} \ll \sum_{x^{1/4} < d < x^{1/2}} \sum_{\substack{d < d' \\ \leqslant \min(dz, x/d)}} y^{\Omega(d')} \frac{x}{dd'} \left(\log \frac{2x}{dd'} \right)^{y-1}.$$

We apply Lemma 45.1 to obtain

$$\sum_{\substack{d < d' \\ \leqslant \min(dz, x/d)}} \frac{y^{\Omega(d')}}{d'} \left(\log \frac{2x}{dd'} \right)^{y-1} \ll \begin{cases} (\log d)^{y-1} (\log z)^{y} & (d > x^{1/2}/z), \\ (\log d)^{y-1} \log z \cdot \left(\log \dfrac{2x}{d^2 z} \right)^{y-1} & \text{(else).} \end{cases} \tag{4.36}$$

When $d \leqslant x^{1/2}/z$, we notice that $\log^{y-1}(2x/d^2 z) \ll \log^{y-1}(2x/d^2)$. So for every $d \in (x^{1/4}, x^{1/2})$ the sum (4.36) is $\ll \log^{y-1} x \cdot \log z \cdot \log^{y-1}(2x/d^2)$. Thus

$$S_2 \ll x \log^{y-1+\alpha} x \sum_{x^{1/4} < d < x^{1/2}} \frac{y^{\Omega(d)}}{d} \left(\log \frac{2x}{d^2} \right)^{y-1}$$
$$\ll x \log^{3y-2+\alpha} x.$$

We put $y = \frac{1}{3}$. If $\xi_1(n) \to \infty$ as $n \to \infty$, we deduce that

$$V_2(n) < \xi_1(n) \cdot 3^{\Omega(n)} \log^{\alpha-1} n \quad \text{p.p.}$$
$$< (\log n)^{\log 3 - 1 + \alpha + \xi(n)/\sqrt{\log\log n}} \quad \text{p.p.} \tag{4.37}$$

by the Hardy–Ramanujan theorem (putting $\xi_1(n) = \log_2 n$ for example).

We introduce the weighted sum

$$V_1'(n) := \sum \{w(n, d): dd'|n, d < d' \leqslant dz, d \leqslant x^{1/4}\} \tag{4.38}$$

where

$$w(n, d) = y^{\Omega(n,d) - \log\log 3d} \quad (y \leqslant 1).$$

Let $\xi_2(x) \to \infty$, $\varepsilon_2 > 0$. By Theorem 11 we have

$$\sup_{\xi_2(x) < d \leqslant x} \left| \frac{\omega(n, d) - \log_2 3d}{\sqrt{(2 \log_2 d \cdot \log_4 d)}} \right| < 1 + \varepsilon_2 \tag{4.39}$$

for almost all integers $n \leqslant x$. Let $\mathscr{A}$ comprise the integers n such that (4.39) holds, and in addition $\Omega(n) - \omega(n) < \varepsilon_2 \sqrt{(2 \log_2 x \cdot \log_4 x)}$, $\Omega(n, \xi_2(x)) < 2 \log_2 \xi_2(x)$. Then $\operatorname{card}(\mathscr{A} \cap [0, x]) = x + o(x)$, and if we further suppose that $2 \log_2 \xi_2(x) < \sqrt{(2 \log_2 x \cdot \log_4 x)}$, we shall have, for $n \in \mathscr{A}$ and *every* d,

$$w(n, d) > w_0 := y^{(1 + 2\varepsilon_2)\sqrt{(2 \log_2 x \cdot \log_4 x)}} \tag{4.40}$$

whence

$$V_1'(n) > w_0 V_1(n) \quad (n \leqslant x, n \in \mathscr{A}).$$

We consider the sum

$$S_1 := \sum_{\substack{n \leqslant x \\ n \in \mathscr{A}}} V_1(n) \leqslant w_0^{-1} \sum_{n \leqslant x} V_1'(n). \tag{4.41}$$

From (4.38) we have

$$\sum_{n \leqslant x} V_1'(n) = \sum_{d \leqslant x^{1/4}} \sum_{d < d' \leqslant dz} \sum_{m \leqslant x/dd'} w(mdd', d)$$

$$\ll \sum_{d \leqslant x^{1/4}} y^{\Omega(d)} (\log 3d)^{-\log y} \sum_{d < d' \leqslant dz} y^{\Omega(d',d)} \frac{x}{dd'} (\log 3d)^{y-1}$$

applying Theorem 01 to the sum over m. Since $d \leqslant x^{1/4}$ we may assume $d \leqslant x/dd'$. To bound the sum over d' we consider two cases according as $d > z$ or not. When $d > z$, we have $d' < d^2$ and $\Omega(d') - \Omega(d', d) \leqslant 1$. So Lemma 45.1 gives

$$\sum_{d < d' \leqslant dz} \frac{y^{\Omega(d',d)}}{d'} \leqslant y^{-1} \sum_{d < d' \leqslant dz} \frac{y^{\Omega(d')}}{d'} \ll_y (\log 3d)^{y-1} \log z.$$

Alternatively when $d \leqslant z$ we drop the condition $d' > d$ and we have

$$\sum_{d' \leqslant dz} \frac{y^{\Omega(d',d)}}{d'} \leqslant \prod_{p \leqslant d} \left(1 - \frac{y}{p}\right)^{-1} \prod_{d < p \leqslant dz} \left(1 - \frac{1}{p}\right)^{-1} \ll (\log 3d)^{y-1} \log z.$$

Hence

$$\sum_{n \leqslant x} V_1'(n) \ll_y x \log^\alpha x \sum_{d \leqslant x^{1/4}} \frac{y^{\Omega(d)}}{d} (\log 3d)^{2y-2-\log y}$$

$$\ll_y x (\log x)^{\alpha + 3y - 2 - \log y}$$

by partial summation, provided $3y - 2 - \log y > 0$. We put $y = \frac{1}{3}$: from (4.41) we have

$$\sum_{\substack{n \leqslant x \\ n \in \mathscr{A}}} V_1(n) \ll w_0^{-1} x (\log x)^{\log 3 - 1 + \alpha}$$

and from this and (4.40),

$$V_1(n) < (\log n)^{\log 3 - 1 + \alpha} B^{\sqrt{(\log_2 n \cdot \log_4 n)}}$$

provided $B > 3^{\sqrt{2}}$ and $\varepsilon_2 = \varepsilon_2(B)$ is sufficiently small. Together with (4.37), this gives the result stated.

4.5 The function $T(n,0)/\tau(n)$

This function is important for the study of Erdős' τ^+-function in the next section. In §4.2 we defined the sequence

$$\mathscr{M}(X) = \{n : T(n,0) > X\tau(n)\} \tag{4.42}$$

and showed that $\mathscr{M}(X)$ possesses logarithmic density $\boldsymbol{\delta}\mathscr{M}(X)$ equal to its lower asymptotic density.

For a general integer sequence $\mathcal{M}$, the upper and lower asymptotic and logarithmic densities satisfy

$$0 \leqslant \underline{\mathbf{d}}\mathcal{M} \leqslant \underline{\boldsymbol{\delta}}\mathcal{M} \leqslant \bar{\boldsymbol{\delta}}\mathcal{M} \leqslant \bar{\mathbf{d}}\mathcal{M} \leqslant 1 \tag{4.43}$$

(Halberstam & Roth (1966), Ch. V, Lemma 1). If the asymptotic density exists so therefore does the logarithmic density, and the two densities are equal.

The question arises whether $\mathbf{d}\mathcal{M}(X)$ exists for every X. We leave this open; it follows from the next theorem that $\mathbf{d}\mathcal{M}(X)$ exists for all but countably many values of X.

Theorem 46. *For $X \geqslant 1$ we have*

$$\bar{\mathbf{d}}\mathcal{M}(X) \ll X^{-1} \log 2X. \tag{4.44}$$

At any continuity point X of $\boldsymbol{\delta}\mathcal{M}(X)$, $\mathbf{d}\mathcal{M}(X)$ exists; that is $T(n,0)/\tau(n)$ has the distribution function

$$F(X) = \begin{cases} 0 & (X < 1), \\ 1 - \boldsymbol{\delta}\mathcal{M}(X+0) & (X \geqslant 1). \end{cases}$$

Later we shall obtain the lower bound $\underline{\mathbf{d}}\mathcal{M}(X) \gg X^{-1} \log^{-1/2} 2X$ (Theorem 47, Cor., (4.61)).

Proof of Theorem 46. We begin with (4.44). We are going to construct a sequence $\mathscr{A}$ such that, for large x,

$$\sum_{\substack{n \leqslant x \\ n \in \mathscr{A}}} T(n,0)/\tau(n) \leqslant c(\mathscr{A})x, \tag{4.45}$$

where $c(\mathscr{A})$ is not too large but nevertheless $\mathbf{d}\mathscr{A}$ is nearly 1. For then

$$\bar{\mathbf{d}}\mathcal{M}(X) \leqslant X^{-1} c(\mathscr{A}) + 1 - \underline{\mathbf{d}}\mathscr{A}. \tag{4.46}$$

We refer the reader to Exercise 10. For each $\kappa > 0$ there exists a $K > 0$ such that, uniformly for $t_0 \geqslant 3$, the sequence $\mathscr{B} = \mathscr{B}(\kappa, t_0)$ of integers n for which

$$\sup_{t \geqslant t_0} \frac{\omega(n,t)}{\log_2 t} > 1 + \kappa \tag{4.47}$$

satisfies $\bar{\mathbf{d}}\mathscr{B} \ll (\log t_0)^{-K}$.

We fix $\kappa = \frac{1}{3}$, and write t_0 in the form $t_0 = \exp(1/\theta_0)$ where θ_0 is small, and at our disposal. $\mathscr{A}$ is to be the complement of $\mathscr{B}$ so that $1 - \underline{\mathbf{d}}\mathscr{A} = \bar{\mathbf{d}}\mathscr{B} \ll \theta_0^K$.

Now we consider the sum

$$\sum_{\substack{n\leqslant x\\ n\in\mathscr{A}}} \tau(n)^{-1} W(n;1,1) \quad (W \text{ as in } (4.18))$$

$$\ll \int_0^\infty |\hat{f}(\theta;1,1)| \sum_{\substack{n\leqslant x\\ n\in\mathscr{A}}} \tau(n)^{-1}|\tau(n,\theta)|^2\,d\theta$$

$$\ll \int_{1/\log x}^{\theta_0} \sum_{\substack{n\leqslant x\\ n\in\mathscr{A}}} \tau(n)^{-1}|\tau(n,\theta)|^2\,d\theta + x\log(2/\theta_0). \tag{4.48}$$

Here we have used the trivial bound $\ll x\log x$ for the inner sum in the range $0\leqslant\theta\leqslant 1/\log x$, and Lemma 30.3 with $\lambda=2$, $y=\frac{1}{2}$ for $\theta>\theta_0$, together with (4.17).

By the definition of $\mathscr{A}$, we have, from (4.47),

$$\omega(n,\exp(1/\theta))\leqslant(1+\kappa)\log(1/\theta) \quad (0<\theta\leqslant\theta_0, n\in\mathscr{A}),$$

and so the integral above does not exceed

$$\int_{1/\log x}^{\theta_0} 2^{(1+\kappa)\log(1/\theta)} \sum_{n\leqslant x} \tau(n)^{-1} 2^{-\omega(n,\exp(1/\theta))}|\tau(n,\theta)|^2\,d\theta$$

$$\ll \frac{x}{\log x}\int_{1/\log x}^{\theta_0} \exp\left\{\frac{1}{4}\sum_{p\leqslant\exp(1/\theta)} p^{-1}|\tau(p,\theta)|^2 + \frac{1}{2}\sum_{\exp(1/\theta)<p\leqslant x} p^{-1}|\tau(p,\theta)|^2\right\}\frac{d\theta}{\theta^\mu},$$

where $\mu=(1+\kappa)\log 2$. The first sum in the exponent is $\leqslant\log(1/\theta)+O(1)$ (estimating crudely), the second is $\leqslant\log_2 x-\log(1/\theta)+O(1)$ by (3.9). So we obtain

$$\ll x\int_{1/\log x}^{\theta_0} \theta^{-\mu}\,d\theta \ll x$$

because $\mu<1$. Since $T(n,0)\leqslant W(n;1,1)$ we now have (4.45) with $c(\mathscr{A})\ll\log(2/\theta_0)$ so that (4.46) becomes

$$\bar{\mathbf{d}}\mathscr{M}(X)\ll X^{-1}\log(2/\theta_0)+\theta_0^K$$

and we set $\theta_0^K=X^{-1}$. This is (4.44).

To prove the second part of the theorem, it will be sufficient after (4.43) to establish

$$\bar{\mathbf{d}}\mathscr{M}(X)\leqslant\boldsymbol{\delta}\mathscr{M}(X) \quad (X\in\mathscr{C}(\boldsymbol{\delta}\mathscr{M}(X))). \tag{4.49}$$

We restrict our attention to integers $n\in\mathscr{A}$, where $\mathscr{A}=\mathbb{Z}^+\setminus(\mathscr{B}(\kappa,t_0)\cup\mathscr{B}_1(l)), \mathscr{B}_1(l):=\{n:\Omega(n)-\omega(n)>l\}$. We write $f(n)=T(n,0)/\tau(n)$.

Then

$$\log x \sum_{\substack{n\leqslant x\\ f(n)>X\\ n\in\mathscr{A}}} 1 \leqslant \sum_{\substack{n\leqslant x\\ f(n)>X\\ n\in\mathscr{A}}} \sum_{p\|n} \log p + \sum_{n\leqslant x}\left(\log x - \sum_{p\|n}\log p\right),$$

and the second term on the right is $\ll x$. Let $\eta>0$, $z=et_0$. We split the first term into three sums S_1, S_2, S_3 in which respectively: $p\leqslant z$; $p>z$ and $f(n/p)>X-\eta$; $p>z$ but $f(n/p)\leqslant X-\eta$. Plainly

$$S_1 \leqslant \sum_{p\leqslant z}\frac{x}{p}\log p \ll x\log t_0. \tag{4.50}$$

Next,

$$S_2 \leqslant \sum_{z<p\leqslant x}\log p \sum_{\substack{m\leqslant x/p\\ f(m)>X-\eta}} 1 \leqslant \sum_{\substack{m\leqslant x\\ f(m)>X-\eta}} \sum_{p\leqslant x/m}\log p$$

$$\leqslant x \sum_{\substack{m\leqslant x\\ f(m)>X-\eta}} \frac{1}{m} + O(x) \leqslant (\delta\mathscr{M}(X-\eta)+o(1))x\log x \tag{4.51}$$

by the Prime Number Theorem and the definition of logarithmic density. Notice that so far we have not used the condition $n\in\mathscr{A}$. We have

$$S_3 = \sum_{\substack{n\leqslant x\\ n\in\mathscr{A}\\ f(n)>X}} \sum_{\substack{p\|n\\ p>z\\ f(n/p)\leqslant X-\eta}} \log p \leqslant \sum_{\substack{n\leqslant x\\ n\in\mathscr{A}}} \sum_{\substack{p\|n\\ p>z\\ f(n)-f(n/p)>\eta}} \log p.$$

Put $n/p=m$. Then

$$T(n,0) = 2\sum_{\substack{d,d'|m\\ |\log(d/d')|\leqslant 1}} 1 + \sum_{\substack{d,d'|n\\ p\|dd'\\ |\log(d/d')|\leqslant 1}} 1 \leqslant 2T(m,0) + \sum_{\substack{\delta\delta'|n\\ p\|\delta\delta'\\ |\log(\delta/\delta')|\leqslant 1\\ (\delta,\delta')=1}} \tau(n/\delta\delta'),$$

whence

$$f(n)-f(m) \leqslant \tau(n)^{-1} \sum_{\substack{\delta\delta'|n\\ P^+(\delta\delta')>z\\ |\log(\delta/\delta')|\leqslant 1}} \tau(n/\delta\delta'),$$

and we denote the function on the right by $f(n,z)$. Evidently the conditions $P^+(\delta\delta')>z$, $|\log(\delta/\delta')|\leqslant 1$ involve $\min(\delta,\delta')>z/e=t_0$, and so for $n\in\mathscr{A}$ and $y\leqslant 1$ we have $f(n,z)\leqslant 2^{l+1}g(n;t_0,y)$ where

$$g(n;t_0,y) := \sum_{\substack{\delta\delta'|n\\ t_0<\delta\leqslant\delta'<e\delta}} 2^{-\Omega(\delta\delta')}y^{\Omega(n,\delta)-(1+\kappa)\log\log 3\delta}; \tag{4.52}$$

hence in S_3 we have $g(n;t_0,y)>2^{-l-1}\eta$ and so

$$S_3 \leqslant 2^{l+1}\eta^{-1}\log x\cdot\sum_{n\leqslant x} g(n;t_0,y). \tag{4.53}$$

Put $\mu=-(1+\kappa)\log y$. We may assume $t_0\geqslant e$ so that $\delta'<\delta^2$ and

$\Omega(\delta', \delta) \geqslant \Omega(\delta') - 1$. Hence, by Theorem 01,

$$\sum_{n \leqslant x} g(n; t_0, y) \ll_y \sum_{t_0 < \delta \leqslant \sqrt{x}} (y/2)^{\Omega(\delta)} (\log 3\delta)^{\mu} \sum_{\delta \leqslant \delta' < \delta e} (y/2)^{\Omega(\delta')} \sum_{m \leqslant x/\delta\delta'} y^{\Omega(m,\delta)}$$

$$\ll \sum_{t_0 < \delta \leqslant \sqrt{x}} x\delta^{-1} (y/2)^{\Omega(\delta)} (\log 3\delta)^{\mu + y/2 - 1} \max\left\{(\log 3\delta)^{y-1}, \left(\log \frac{3x}{\delta^2}\right)^{y-1}\right\}$$

$$\ll_y x(\log t_0)^{\mu + 2y - 2}$$

provided this last exponent is negative. This is the case if $\kappa \in (0, 1)$, $y = (1 + \kappa)/2$: (4.53) yields

$$S_3 \ll 2^l \eta^{-1} (\log t_0)^{-\beta} x \log x$$

with $\beta = Q(1 + \kappa) - (1 + \kappa)\log 2 + 1$. We put this together with (4.50), (4.51), and let $x \to \infty$ to obtain

$$\bar{\mathbf{d}}(\mathscr{M}(X) \cap \mathscr{A}) \leqslant \boldsymbol{\delta}\mathscr{M}(X - \eta) + 2^l \eta^{-1} (\log t_0)^{-\beta}.$$

We also have

$$\bar{\mathbf{d}}(\mathscr{B}(\kappa, t_0) \cup \mathscr{B}_1(l)) \ll (\log t_0)^{-K} + l^{-1},$$

(since the average order of $\Omega(n) - \omega(n)$ is a constant). We let $t_0 \to \infty$, $l \to \infty$, $\eta \to 0$: (4.49) follows.

4.6 Erdős' function $\tau^+(n)$

For $n \in \mathbb{Z}^+$ and real u, let $\varepsilon(n, u) = 1$ or 0 according as n has, or has not, a divisor $d \in (2^u, 2^{u+1}]$. We define

$$\tau^+(n) := \sum_{k \in \mathbb{Z}} \varepsilon(n, k)$$
$$= \text{card}\,\{k \in \mathbb{Z}: \exists d: d|n, 2^k < d \leqslant 2^{k+1}\}. \tag{4.54}$$

Plainly

$$\Omega(n) + 1 \leqslant \tau^+(n) \leqslant \min\left\{\tau(n), \left[\frac{\log n}{\log 2}\right] + 1\right\}, \tag{4.55}$$

the sequence $n = 2^m$ showing that all three inequalities are best possible.

If $\tau^+(n) < \tau(n)$, then n has divisors d, d' such that $d < d' < 2d$. A number such as 15 shows that the converse is false.

Lemma 47.1. *For every n,*

$$\tau(n)^2 \leqslant \tau^+(n)\, T(n, 0) \tag{4.56}$$

Proof. Put

$$v(n, k) := \text{card}\,\{d: d|n, 2^k < d \leqslant 2^{k+1}\}$$

Then

$$\tau(n) = \sum_k \nu(n,k) = \sum_k \varepsilon(n,k)\nu(n,k),$$

$$\tau^+(n) = \sum_k \varepsilon(n,k) = \sum_k \varepsilon(n,k)^2.$$

Cauchy's inequality gives

$$\tau(n)^2 \leqslant \tau^+(n) \sum_k \nu(n,k)^2$$

and we have

$$\sum_k \nu(n,k)^2 = \sum_{d|n} \sum_{d'|n} \operatorname{card}\{k: 2^k < d, d' \leqslant 2^{k+1}\}$$
$$\leqslant \operatorname{card}\{d, d' : d, d'|n, |\log d'/d| < \log 2\} \leqslant T(n, 0).$$

Theorem 47. *Let*

$$\mathcal{N}(\alpha) := \{n: \tau^+(n) < \alpha\tau(n)\}$$

Then, for $0 < \alpha \leqslant 1$,

$$\bar{\mathbf{d}}\mathcal{N}(\alpha) \ll \alpha \log(2/\alpha) \tag{4.57}$$

$$\underline{\mathbf{d}}\mathcal{N}(\alpha) \gg \alpha \log^{-1/2}(2/\alpha). \tag{4.58}$$

Proof. We begin with (4.57). Let $\tau^+(n) < \alpha\tau(n)$. By Lemma 47.1, $T(n,0) > \alpha^{-1}\tau(n)$, whence $\mathcal{N}(\alpha) \subseteq \mathcal{M}(1/\alpha)$ and the result is a corollary of Theorem 46. It remains to prove (4.58).

We introduce the supplementary function

$$\tau^*(n) := \max_{0 \leqslant \lambda < 1} \operatorname{card}\{k: \exists d: d|n, 2^{k+\lambda} \leqslant d < 2^{k+\lambda+1}\}.$$

Clearly $\tau^+(n) \leqslant \tau^*(n)$, moreover τ^* satisfies

$$\tau^*(mn) \leqslant \tau(m)\tau^*(n), \quad ((m,n) = 1), \tag{4.59}$$

which τ^+ may not: witness $m = 19$, $n = 35$. We set

$$n(w) := \prod_{\substack{p^\nu \| n \\ p \leqslant w}} p^\nu \quad (w \geqslant 3)$$

where w will depend on α, and we consider the sequence $\mathcal{A}$ of integers n for which

$$\omega(n(w)) \geqslant 2 \log_2 w, \quad n(w) \leqslant w. \tag{4.60}$$

We have

$$\underline{\mathbf{d}}\mathcal{A} = \prod_{p \leqslant w} \left(1 - \frac{1}{p}\right) \sum_{\substack{a \leqslant w \\ \omega(a) \geqslant 2\log_2 w}} a^{-1}.$$

Put $k := [2 \log\log w] + 1$. By the classical estimate of Sathe and Selberg (see

Exercise 48) we deduce that

$$\begin{aligned}\underline{\mathbf{d}}\mathscr{A} &\gg (\log w)^{-1} \sum_{a \leqslant w, \omega(a)=k} a^{-1} \\ &\gg (\log w)^{-1}(\log_2 w + O(1))^k/k! \\ &\gg (\log w)^{-Q(2)}(\log_2 w)^{-1/2}.\end{aligned}$$

Now, by (4.59) and (4.60), for $n \in \mathscr{A}$,

$$\begin{aligned}\tau^*(n)/\tau(n) &\leqslant \tau^*(n(w))/\tau(n(w)) \leqslant ([(\log w)/(\log 2)] + 1)2^{-2\log_2 w} \\ &< 3(\log w)^{1-\log 4}.\end{aligned}$$

We fix $w = w(\alpha)$ so that the right-hand side equals α. Since $Q(2) = \log 4 - 1$ and $\tau^+(n) \leqslant \tau^*(n)$, the result follows.

Corollary.

$$\mathbf{d}\mathscr{M}(X) \gg X^{-1}\log^{-1/2}(2X) \quad (X \geqslant 1). \tag{4.61}$$

At one time Erdős was almost certain that $\tau^+(n) = o(\tau(n))$ p.p., and thought that his old conjecture could perhaps be proved in this way. But this was shown to be false by Erdős and Tenenbaum (1981) who obtained a slightly weaker form of (4.57). They suggested that $\tau^+(n)/\tau(n)$ might have a distribution function, which is continuous and strictly increasing on $[0, 1]$. We shall establish here the first part of this conjecture.

One of the intrinsic difficulties in the study of $\tau^+(n)$ arises essentially from the special rôle played by the powers of 2 in the definition. It will be useful to first introduce a modification which avoids this drawback.

Put, for $0 \leqslant \lambda < 1$,

$$\tau^+(n, \lambda) = \sum_{k \in \mathbb{Z}} \varepsilon(n, k + \lambda)$$

so that

$$\tau^+(n) = \tau^+(n, 0), \quad \tau^*(n) = \sup_{0 \leqslant \lambda < 1} \tau^+(n, \lambda).$$

The reader may check by means of examples that $\tau^+(n, \lambda)$ is quite sensitive to changes in λ, varying by a factor as much as 2. It is natural to average over λ and, accordingly, we write

$$\tau^\dagger(n) := \int_0^1 \tau^+(n, \lambda)\, \mathrm{d}\lambda = \int_{-\infty}^{+\infty} \varepsilon(n, u)\, \mathrm{d}u.$$

For $u - 1 \leqslant v \leqslant u$, we have $\varepsilon(n, u) \leqslant \varepsilon(n, v) + \varepsilon(n, v + 1)$, and this implies that for all n

$$\tfrac{1}{2}\tau^*(n) \leqslant \tau^+(n), \quad \tau^\dagger(n) \leqslant \tau^*(n).$$

This seems hard to improve, even only for large n. However we shall show the following.

Theorem 48. *We have*

$$\tau^+(n) \sim \tau^\dagger(n) \sim \tau^*(n), \text{ p.p.}$$

Thus the distribution problem mentioned above may be tackled by dealing with any of the three functions. Naturally, $\tau^\dagger(n)$ is the most convenient.

Theorem 49. *The function $\tau^\dagger(n)/\tau(n)$ has a distribution function.*

Corollary. *The above statement also holds for the functions $\tau^+(n)/\tau(n)$, and $\tau^*(n)/\tau(n)$.*

The proofs of both theorems rely on the following Lemma, established in Erdős–Tenenbaum (1983), pp. 135–7.

Lemma 48.1. *For each $\varepsilon > 0$ there exists a sequence $\mathscr{A}(\varepsilon)$ with $\bar{\mathbf{d}}\mathscr{A}(\varepsilon) < \varepsilon$, and such that if $n \in \mathbb{Z}^+ \backslash \mathscr{A}(\varepsilon)$ then we may write $n = ab$, where $a < P^-(b)$, $\tau(a) > \varepsilon^{-1}$, and b possesses a sequence of at least $(1-\varepsilon)\tau(b)$ divisors $b_1 < b_2 < \cdots < b_s$ with the property that b_i is the only divisor of b in $[b_i/2a, 2ab_i]$ $(1 \leqslant i \leqslant s)$.*

Moreover, the function

$$a(n,\varepsilon) := \begin{cases} a & (n \notin \mathscr{A}(\varepsilon), n = ab) \\ n & (n \in \mathscr{A}(\varepsilon)) \end{cases}$$

satisfies conditions (i) and (iii) of Lemma A2.

This may seem rather technical, but the first part is a standard device worth bearing in mind. Broadly, it states that it is possible to split the prime factors of an integer so that most ratios of consecutive divisors made up of the large primes exceed the product of all the small primes. It is relevant for problems concerning the two orders of the divisors.

Proof of Theorem 48. We show that $\tau^+(n) \sim \tau^\dagger(n)$, p.p. An obvious extension of the argument would yield that for almost all n, $\tau^+(n,\lambda) \sim \tau^\dagger(n)$ holds uniformly in $\lambda \in [0, 1)$, whence the required assertion concerning $\tau^*(n)$.

In this section only we employ the notation $\operatorname{Log} v := (\log v)/(\log 2)$. With $a(n,\varepsilon)$ defined as in Lemma 48.1, we put $b(n,\varepsilon) := n/a(n,\varepsilon)$ and let $D(n,\varepsilon)$ denote the discrepancy of the numbers $\operatorname{Log}\beta$, $\beta | b(n,\varepsilon)$.

Lemma 48.2. *We have*

$$\lim_{\varepsilon \to 0+} \bar{\mathbf{d}}\{n: D(n,\varepsilon) > \varepsilon\} = 0.$$

Proof. Let $\eta > 0$ be given. We show that for suitable $\varepsilon(\eta) > 0$ the upper density above is $\leqslant \eta$ whenever $0 < \varepsilon \leqslant \varepsilon(\eta)$. Let $\mathscr{B}(\varepsilon, T)$ comprise the integers n such that $a(n,\varepsilon) \leqslant T$. By the last statement of Lemma 48.1, we have

$$\bar{\mathbf{d}}(\mathbb{Z}^+ \backslash \mathscr{B}(\varepsilon, T)) \leqslant \eta$$

for $\varepsilon \leqslant \varepsilon(\eta)$ and $T = T(\varepsilon, \eta)$ suitably chosen. Now observe that for any fixed, real $\theta \neq 0$ we have by a straightforward application of Theorem 01 and Lemma 30.2

$$\sum_{\substack{n \leqslant x \\ n \in \mathscr{B}(\varepsilon,T)}} |\tau(b(n,\varepsilon),\theta)|/\tau(b(n,\varepsilon)) \leqslant \sum_{a \leqslant T} \sum_{\substack{b \leqslant x/a \\ P^-(b) > a}} |\tau(b,\theta)|/\tau(b)$$
$$\ll_{\varepsilon,\theta} x(\log x)^{2/\pi - 1} = o_{\varepsilon,\theta}(x).$$

In view of the Erdős–Turán inequality (see Kuipers & Niederreiter (1974), Theorem 2.5)

$$D(n,\varepsilon) \ll \frac{1}{m} + \sum_{h=1}^{m} \frac{|\tau(b(n,\varepsilon), h/\log 2)|}{h\tau(b(n,\varepsilon))} \quad (m \geqslant 1),$$

this shows that $D(n,\varepsilon) = o(1)$ for almost all n in $\mathscr{B}(\varepsilon, T)$. This is all we need.

For our next lemma, we introduce some extra notation. We write $B_1(t)$: $= \{t\} - \frac{1}{2}$ the first Bernouilli function. For $n \geqslant 1$, we say that a divisor d of n is *isolated* if d is the only divisor of n in $[d/2, 2d]$. When $[d/2, d)$ (resp. $(d, 2d]$) contains a divisor of n but $(d, 2d]$ (resp. $[d/2, d)$) does not, we call d *right-isolated* (resp. *left-isolated*). We denote respectively by $\mathscr{I}(n)$, $\mathscr{R}(n)$ and $\mathscr{L}(n)$ the sets of isolated, right-isolated and left-isolated divisors of n.

Lemma 48.3. *We have for all n*

$$\tau^\dagger(n) - \tau^+(n) = \sum_{d \in \mathscr{R}(n)} B_1(\operatorname{Log} d) - \sum_{d \in \mathscr{L}(n)} B_1(\operatorname{Log} d). \tag{4.62}$$

Proof. Let $1 = d_1 < d_2 < \cdots < d_\tau = n$ denote the ordered sequence of the divisors of n and set $\mathscr{R}(n) = \{d_{j_1}, \ldots, d_{j_r}\}$, $\mathscr{L}(n) = \{d_{i_1}, \ldots, d_{i_l}\}$. Then $r = l$ and $i_1 < j_1 < i_2 < j_2 < \cdots < i_r < j_r$. Moreover, a simple computation gives that

$$\tau^+(n) = \operatorname{card} \mathscr{I}(n) + \sum_{1 \leqslant m \leqslant r} (1 + [\operatorname{Log} d_{j_m}] - [\operatorname{Log} d_{i_m}]),$$
$$\tau^\dagger(n) = \operatorname{card} \mathscr{I}(n) + \sum_{1 \leqslant m \leqslant r} (1 + \operatorname{Log} d_{j_m} - \operatorname{Log} d_{i_m}).$$

This yields the asserted formula.

We are now able to complete the proof of Theorem 48. By (4.57), it is sufficient to show that

$$\tau^+(n) - \tau^\dagger(n) = o(\tau(n)), \text{ p.p.}$$

Let $R(n)$ denote the first sum on the right of (4.62). We set out to prove that for arbitrary $\varepsilon, \eta > 0$,

$$\bar{\mathbf{d}}\{n: |R(n)| > \varepsilon\tau(n)\} \leqslant \eta.$$

Of course we can assume here that $\varepsilon \leqslant \varepsilon(\eta)$. By Lemmas 48.1, 48.2, we may hence restrict our attention to the case $n \notin \mathscr{A}(\varepsilon)$, $D(n, \varepsilon) \leqslant \varepsilon$. Then, we have

$$\begin{aligned} R(n) &= \sum_{d \in \mathscr{R}(n)} B_1(\operatorname{Log} d) = \sum_{\alpha | a} \sum_{\substack{\beta | b \\ \alpha\beta \in \mathscr{R}(n)}} B_1(\operatorname{Log}(\alpha\beta)) \\ &= \sum_{\alpha | a} \sum_{\substack{1 \leqslant i \leqslant s \\ \alpha b_i \in \mathscr{R}(n)}} B_1(\operatorname{Log}(\alpha b_i)) + O(\varepsilon\tau(n)) \end{aligned}$$

where the b_i are as in Lemma 48.1. Now for any $d = \alpha b_i$, all the possible divisors d' of n in $[d/2, 2d]$ are of the form $\alpha' b_i$ with $\alpha' | a$. Hence $\alpha b_i \in \mathscr{R}(n)$ if, and only if, $\alpha \in \mathscr{R}(a)$. It follows that

$$\begin{aligned} R(n) &= \sum_{\substack{\alpha | a \\ \alpha \in \mathscr{R}(a)}} \sum_{1 \leqslant i \leqslant s} B_1(\operatorname{Log}(\alpha b_i)) + O(\varepsilon\tau(n)) \\ &= \sum_{\substack{\alpha | a \\ \alpha \in \mathscr{R}(a)}} \sum_{\beta | b} B_1(\operatorname{Log}(\alpha\beta)) + O(\varepsilon\tau(n)). \end{aligned}$$

From Koksma's inequality (see e.g. Kuipers and Niederreiter (1974), Theorem 5.1) the β-sum above has modulus at most $D(n, \varepsilon)\tau(b) \leqslant \varepsilon\tau(b)$, whence $R(n) \ll \varepsilon\tau(n)$. The second sum in (4.62) may be treated similarly. This completes the proof.

Proof of Theorem 49. We first notice that for $(m, n) = 1$

$$\begin{aligned} \tau^\dagger(mn) &= \int_{-\infty}^{+\infty} \varepsilon(mn, u)\,du \leqslant \int_{-\infty}^{+\infty} \sum_{d | m} \varepsilon(n, u - \operatorname{Log} d)\,du \\ &\leqslant \tau(m)\tau^\dagger(n) \end{aligned}$$

so that

$$\frac{\tau^\dagger(mn)}{\tau(mn)} \leqslant \frac{\tau^\dagger(n)}{\tau(n)} \quad ((m, n) = 1) \tag{4.63}$$

In view of this and Lemmas 48.1 and A2 it will therefore be sufficient to establish that for all n

$$\frac{\tau^\dagger(n)}{\tau(n)} - \frac{\tau^\dagger(a(n, \varepsilon))}{\tau(a(n, \varepsilon))} \geqslant -\varepsilon,$$

and we may assume $n \in \mathbb{Z}^+ \backslash \mathscr{A}(\varepsilon)$, $n = ab$. With $\{b_i\}$ as in Lemma 48.1, evidently

$$\tau^\dagger(ab) \geqslant \sum_{i=1}^{s} \int_{\operatorname{Log}(b_i/a)}^{\operatorname{Log} ab_i} \varepsilon(ab, u)\,du$$

(the ranges not overlapping), whence

$$\tau^{\dagger}(ab) \geqslant \sum_{i=1}^{s} \int_{-\operatorname{Log} a}^{\operatorname{Log} a} \varepsilon(ab, u + \operatorname{Log} b_i)\, du \geqslant \sum_{i=1}^{s} \int_{-\operatorname{Log} a}^{\operatorname{Log} a} \varepsilon(a, v)\, dv$$

(since we may demand that the divisor of ab should be a multiple of b_i). For $v \notin [-1, \operatorname{Log} a)$, $\varepsilon(a, v) = 0$, and we may assume that $\operatorname{Log} a > 1$. Therefore the integral above equals $\tau^{\dagger}(a)$ and we have

$$\tau^{\dagger}(ab) \geqslant (1 - \varepsilon)\tau(b)\tau^{\dagger}(a).$$

The desired inequality follows.

The average order of $\tau^{+}(n)$ presents an open problem. The following is known.

Theorem 410. *For a suitable constant c, we have*

$$x(\log x)^{1-\delta} \exp\{-c\sqrt{(\log_2 x \cdot \log_3 x)}\} \ll \sum_{n \leqslant x} \tau^{+}(n)$$
$$\ll x(\log x)^{1-\delta}(\log_2 x)^{-1/2} \quad (x \geqslant 16),$$

where $\delta = .08607\ldots$ *is as in* (2.3).

We recall the definition of $H(x, y, z)$ from Chapter 2: this is the number of integers $n \leqslant x$ having at least one divisor in $(y, z]$. Evidently

$$\sum_{n \leqslant x} \tau^{+}(n) = \sum_{k=-1}^{\infty} H(x, 2^k, 2^{k+1}).$$

By the symmetry of the divisors of n about $\sqrt{n}$, we see that those $k < [(\log x)/(\log 4)]$ contribute together at least one half of the initial sum. Appealing then to Theorem 21 (iii), we obtain the asserted upper estimate. For the lower bound, we restrict to the range $\frac{1}{2}\log x < k < [(\log x)/(\log 4)]$ and apply (2.11) with $y = 2^k$, $z = 2^{k+1} \leqslant \sqrt{x}$, $u = 1/k$.

Presumably the upper bound above is nearer the truth; the lower bound derivation is somewhat inefficient.

4.7 Hooley's function Δ(n)

This is the greatest number of divisors of n contained in any interval of logarithmic length 1, that is

$$\Delta(n) = \max_{z} \operatorname{card}\{d : d | n, z < d \leqslant ez\}. \tag{4.64}$$

This function and its generalization $\Delta_r(n)$ receive extensive attention in Chapter 6 (where $\Delta_r(n)$ is defined) and Chapter 7, partially because of the interest arising from the applications to other branches of number theory envisaged by Hooley, and carried through by Hooley (1979) and Vaughan (1986, 1986a) for particular problems. However the applications rather focus our attention on the average orders of $\Delta(n)$ and $\Delta_r(n)$, and the normal order of $\Delta(n)$ is also important: it is obviously related to Erdős' conjecture concerning divisors $d < d' < 2d$ and is tackled in Chapter 5.

We restrict ourselves here to partial solutions of two extremal problems: suppose that we are given the value of $T(n,0)$ or $U(n,0)$ but no other information about n. What are the best upper bounds for $\Delta(n)$?

It is clear that

$$\Delta(n) \leqslant T(n,0)^{1/2}; \tag{4.65}$$

moreover this is nearly best possible: let $h \in \mathbb{Z}^+$, $11 \leqslant p_1 < p_2 < \cdots < p_{2h} < (1 + 1/2h)p_1$ and $n = p_1 p_2 \ldots p_{2h}$. For $d, d' | n$, we have $|\log(d/d')| < 1$ if and only if $\omega(d) = \omega(d')$, whence

$$\Delta(n) = \binom{2h}{h}, \quad T(n,0) = \sum_{j=0}^{2h} \binom{2h}{j}^2 = \binom{4h}{2h},$$

and so $\Delta(n)^2 \gg T(n,0)/\sqrt{\log T(n,0)}$. It is an open problem whether or not $\Delta(n) = o(\sqrt{T(n,0)})$ $(T(n,0) \to \infty)$.

In this example we also have

$$U(n,0) = \sum_{j=0}^{h} \binom{2h}{2j}\binom{2j}{j} \sim ch^{-1/2} \cdot 3^{2h}, \tag{4.66}$$

whence $\Delta(n)^\beta \gg U(n,0)/(\log U(n,0))^{(\beta-1)/2}$ where $\beta = \log 3/\log 2$. In fact we can have $\Delta(n)^{1+\varepsilon} > U(n,0)$. Let $n = (p_1 p_2 \ldots p_k)^\alpha$. Then $U(n,0) \leqslant (2\alpha + 1)^k$, and, by taking the primes close together, we can arrange that $\Delta(n) \geqslant \tau(n)/(\Omega(n) + 1) = (\alpha + 1)^k/(\alpha k + 1)$. However, we do have

$$\Delta(n) \leqslant U(n,0)^{1/\beta} \text{ for squarefree } n. \tag{4.67}$$

We need the following general result.

Theorem 411. *Let n be squarefree, and R, S be (not necessarily disjoint) sets of respectively $|R|$, $|S|$ divisors of n. Then there are at least $(|R||S|)^{\beta/2}$ distinct ratios of the form r/s, $r \in R$, $s \in S$.*

This is best possible: let $R = S = \{d: d|m\}$ where m is any divisor of n. Then $|R| = |S| = 2^{\omega(m)}$, and there are $3^{\omega(m)}$ ratios.

Before the proof, we deduce (4.67). Let $\Delta(n) = t$, and $z_0 < d^{(1)} < d^{(2)} < \cdots < d^{(t)} \leqslant ez_0$. Put $R = S = \{d^{(1)}, d^{(2)}, \ldots, d^{(t)}\}$. By the theorem, there are at least t^β distinct ratios $d^{(i)}/d^{(j)}$ and we may write $d^{(i)}/d^{(j)} =: \delta_{ij}/\delta'_{ij}$ in its lowest terms. The pair δ_{ij}, δ'_{ij} contributes to $U(n,0)$ and plainly the pairs are distinct. Hence $U(n,0) \geqslant t^\beta$ as required.

Lemma 411.1 (Woodall (1977)). *Let R and S be (not necessarily disjoint) sets of vertices of a k-dimensional hypercube, and M comprise all vectors of the form $\mathbf{m} = \mathbf{r} + \mathbf{s}$, $\mathbf{r} \in R$, $\mathbf{s} \in S$. Then $|M| \geqslant (|R||S|)^{\beta/2}$.*

For the proof we refer the reader to Woodall's paper.

Proof of Theorem 411. Let $n = p_1 p_2 \ldots p_k$ and K be the hypercube with vertex set $\{0, 1\}^k$. With each divisor $r = p_1^{\varepsilon_1} p_2^{\varepsilon_2} \ldots p_k^{\varepsilon_k} \in R$ we associate the vector $\mathbf{r} = (\varepsilon_1, \varepsilon_2, \ldots, \varepsilon_k)'$, which is a vertex of K, and with each divisor $s = p_1^{\delta_1} p_2^{\delta_2} \ldots p_k^{\delta_k} \in S$ we associate the vector $\mathbf{s} = (1 - \delta_1, 1 - \delta_2, \ldots, 1 - \delta_k)'$. By the lemma, there are at least $(|R||S|)^{\beta/2}$ distinct vectors $\mathbf{m} = \mathbf{r} + \mathbf{s}$, and $\mathbf{m} = (1 + \mu_1, 1 + \mu_2, \ldots, 1 + \mu_k)'$ where $\mu_i = -1$, 0 or 1 $(1 \leqslant i \leqslant k)$. Each $\mathbf{m}$ gives rise to the ratio $p_1^{\mu_1} p_2^{\mu_2} \ldots p_k^{\mu_k}$, and these ratios are distinct and of the form r/s, $r \in R$, $s \in S$. This completes the proof.

Notes on Chapter 4

§4.3. The method of proof used here is not applicable when $y>\frac{1}{2}$. As we remarked, in this case the exponents of $\log x$ in Theorem 41 become $4y-2+\alpha$ and $3y-2+\alpha$ (the formula involving $U(n,\alpha)$ is given, with a sketch proof, in Hall & Tenenbaum (1981)). Now let $I(\beta)=\{\theta:(\log x)^{-\beta}<|\theta|<\exp\sqrt{\log x}\}$, $\alpha<\beta<1$, $y>\frac{1}{2}$. We have (for $h \doteqdot (\log x)^{\alpha}$)

$$\int_{I(\beta)} \hat{f}(\theta;h,\eta)\sum_{n\leqslant x}|\tau(n,\theta)|^2 y^{\Omega(n)}\,d\theta \ll x(\log x)^{(1+\beta)(2y-1)+\alpha}$$

which is negligible, so that the main term comes essentially from the range $|\theta|<1/\log x$ where the results of Chapter 3 do not apply.

§4.5. The idea needed for the proof of (4.45) is due to A. Hildebrand, who proved that if $\xi(n)\to\infty$ then $T(n,0)<\xi(n)\tau(n)$ p.p., using essentially Theorem 11 rather than Exercise 10 to construct a suitable $\mathscr{A}$ with $\mathbf{d}\mathscr{A}=1$.

§4.7. In analogy to the definition of a *highly composite number* (Ramanujan (1915a)), Erdős & Nicolas (1976) call n Δ-abundant if $\Delta(n)>\max\{\Delta(m): m<n\}$. They show that if n is Δ-abundant then

$$\Delta(n) \doteqdot \tau(n)/\sqrt{(\log n\cdot\log_2 n)}.$$

Exercises on Chapter 4

40. Prove as simply as possible, using Fourier analysis and results from Chapter 3, that, for $0 \leqslant \alpha \leqslant 1$, $\frac{1}{2} < y \leqslant 1$,

$$\sum_{n \leqslant x} T(n, \alpha) y^{\Omega(n)} \asymp x (\log x)^{4y - 2 + \alpha}.$$

41. Combine Exercise 40 with Hölder's inequality to obtain for $p < 2$ that

$$\sum_{n \leqslant x} T(n, \alpha)^{1/p} \ll x (\log x)^{4^{1/p} - 1 - (1 - \alpha)/p}.$$

Show that, if $\alpha < 1$ and $p > p_0(\alpha)$, this is false.

42. Deduce from the proposition that $\tau(n)$ has a normal order that if $\xi(n) \to \infty$ then $\omega(n) - \omega(n-1) > -\xi(n)$ p.p. Why is this false? [Hint: cf. Kubilius (1964) Chapter V.]

43. Let $S \subseteq [1, x]$ be a set of integers such that $x - |S| \ll x/(\log x)^{Q(\alpha)}$ $(Q(\alpha) = \alpha \log \alpha - \alpha + 1)$, where $0 < \alpha < 1$. Prove that

$$\sum_{n \in S} 2^{\omega(n)} \gg_\varepsilon x (\log x)^{\alpha \log 2 - Q(\alpha) - \varepsilon}.$$

44. Combine Exercise 43 with Theorem 47 to prove that

$$\sum_{n \leqslant x} \tau^+(n) \gg_\varepsilon x (\log x)^{2(\sqrt{2} - 1) - \varepsilon}.$$

Examine why this result is weaker than Theorem 410.

45. Deduce from (4.65) that

$$\sum_{n \leqslant x} \Delta(n) \ll x \sqrt{(\log x \cdot \log_2 x)}.$$

46. Let $\Delta(n; u) := \tau(n, e^u, e^{u+1})$. Show that

$$\int_{-\infty}^{\infty} \Delta(n; u)^2 \, du \leqslant T(n, 0).$$

47. Show that, for $m, n \in \mathbb{Z}^+$,

$$\Delta(mn) \leqslant \max_u \sum_{d \mid m} \Delta(n; u - \log d) \qquad (1)$$

and deduce

(i) $\Delta(mn) \leqslant \tau(m) \Delta(n)$,

(ii) $\Delta(mn)^2 \leqslant 4 T(m, 0) T(n, 0)$.

[Hint: majorize the right-hand side of (1) by a suitable convolution integral. Apply Exercise 46.]

48. Put $F(z) = \Gamma(z+1)^{-1}\prod_p (1-1/p)^z(1+z/(p-1))$ and assume that Selberg's formula (1954)

$$\sum_{n \leqslant x} z^{\omega(n)} = zF(z)x(\log x)^{z-1} + O_B(x(\log x)^{z-2})$$

holds for $x \geqslant 2$, $z \in \mathbb{C}$, $|z| \leqslant B$. Deduce the estimate

$$\sum_{\substack{n \leqslant x \\ \omega(n)=k}} 1 \sim F\left(\frac{k-1}{\log_2 x}\right)\frac{x}{\log x}\frac{(\log_2 x)^{k-1}}{(k-1)!}$$

uniformly for $x \to \infty$, $k \leqslant B \log_2 x$.

5
Erdős' conjecture

5.1 Introduction and results

Historically, Erdős' conjecture first appeared as an epiphenomenon. In the thirties, the main interest in this circle of ideas was in the structure of sequences rather than the structure of numbers – and the focus of attention was the concept of a set of multiples. After Chowla, Davenport and Erdős had proved (independently) in 1934 that the abundant (*) numbers have asymptotic density, a deeper understanding of the situation was sought, possibly in the form of a general theorem stating that every set of multiples had a density. However Besicovitch disproved this by an explicit construction (cf. Chapter 2) based on the observation that for fixed, large y it is unlikely that an integer has a divisor in $(y, 2y]$.

Certainly not the least of Erdős' achievements was to have recognized that the occurrence of such questions about divisors in 'short' intervals was by no means a matter of chance. The attempt to answer these questions has proved to be exactly the way to improve the model of normal integers as based purely on the double exponential growth of the prime factors and in accordance (as we saw in Chapter 1) with the fact that the expected value for $d_j(n)$ is $\exp(j^{1/\log 2})$ $(1 \leqslant j \leqslant \tau(n))$.

Consider the function

$$\Delta(n; y) := \operatorname{card}\{d : d|n, e^y < d \leqslant e^{y+1}\}.$$

In the above model, this would just take the values 0 and 1 – and Besicovitch's result also points in this direction. But Erdős conjectured that almost all integers have two divisors d, d' such that

$$d < d' \leqslant 2d$$

and indeed that $\Delta(n) = \max \Delta(n; y) \to \infty$ p.p. Such a result clearly does not follow from the results on the normal distribution of prime factors set out in Chapter 1. We have here a good example of how Erdős' apparently purely speculative questions help to further our knowledge of the structure of integers. In the light of the rest of the theory, this very simple conjecture is

(*) An abundant number has $\sigma(n) = \sum\{d : d|n\} > 2n$.

revealed as both deep and delicate, reflecting the crudity of the available model. The conjecture can be put in a probabilistic framework where it raises the fundamental problem of the estimation of the *concentration* of a sum of random variables – see the Notes.

The sequence of those n with Erdős' required property is easily seen to be a set of multiples. In 1948 Erdős gave a criterion for such a sequence to have asymptotic density which applied, positively, in the present case. The question as to whether or not this density equals 1 was to remain open for another 35 years!

The heuristic argument which let Erdős to formulate his conjecture was based on the assumption that the numbers $\{\log(d/d'): d, d'|n, (d,d')=1\}$ are evenly distributed in $[-\log n, \log n]$. As we saw in §4.3 this suggests that

$$U(n,0) := \text{card}\{d, d' : dd'|n, (d,d')=1, |\log(d/d')| \leqslant 1\}$$
$$= (\log n)^{\log 3 - 1 + o(1)} \quad \text{p.p.},$$

and we proved in Theorem 45 that the upper bound implied by this formula is valid. We need the corresponding lower bound. Having various applications in mind, and to avoid repetition of a similar underlying argument in different contexts, we make a generalization in two directions. First, we introduce the distribution function

$$\nabla(n,t) := \text{card}\{d, d'|n: (d,d')=1, |\log(d/d')| \leqslant t\}$$

and second, we consider for each n the partial product

$$n(u,v) := \prod_{\substack{p^\nu \| n \\ u<p\leqslant v}} p^\nu$$

where u and v are free parameters. The usual order of either $\omega(n(u,v))$ or $\Omega(n(u,v))$ is

$$w := \log\left(\frac{\log v}{\log u}\right) \quad (u<v), \tag{5.1}$$

whence the number of pairs $d, d'|n(u,v), (d,d')=1$, is as expected, 3^w. Moreover by Theorem 07 we may also expect $\log n(u,v)$ to be not much larger than $\log v$. This suggests that the average distance between numbers of the form $\log(d/d')$, with $d, d'|n(u,v)$ should have order of magnitude about

$$h := 3^{-w}\log v, \tag{5.2}$$

and further, assuming reasonably even distribution in $[-\log v, \log v]$, we are led to expect that

$$\nabla(n(u,v),t) \doteqdot h^{-1}t \quad (0 \leqslant t \leqslant \log v).$$

The following two theorems, which are the main content of this chapter, show that this very heuristic argument is not too unrealistic.

Theorem 50. *Let u, v, x be such that*

$$2 \leqslant u^3 \leqslant v \leqslant x \tag{5.3}$$

and w and h be as in (5.1) *and* (5.2). *Then, for all* $t \geqslant 0$ *and* $\xi \in (0, \sqrt{w}]$, *we have*

$$\nabla(n(u,v),t) \leqslant 1 + h^{-1} t \exp(\xi\sqrt{w}) \tag{5.4}$$

for all except $\ll xw\exp(-\xi^2/11)$ *of the integers* $n \leqslant x$.

Theorem 51. *Let u, v, x, w, h and* ξ *be as above, and*

$$\beta := 1 - (1 + \log_2 3)/\log 3 = 0.004\,15\ldots. \tag{5.5}$$

Then for all $t \in [0, \log v]$ *we have*

$$\nabla(n(u,v),t) > h^{-1} t \exp(-\xi\sqrt{w}) \tag{5.6}$$

for all except

$$\ll x\{\exp(-\xi^2/50) + \xi^{-\beta} w^{-\beta/2} (\log w)^{4\beta}\}$$

of the integers $n \leqslant x$.

We now derive, by various specializations of the parameters, the following consequences. First things first: we have the following.

Theorem 52. *Almost all integers have two divisors* d, d' *such that* $d < d' \leqslant 2d$; *moreover the number of integers* $n \leqslant x$ *without such divisors is*

$$\ll x(\log_2 x)^{-\beta}(\log_3 x)^{4\beta}.$$

Proof. We apply Theorem 51 with $u = \frac{3}{2}$, $v = x$, $t = \log 2$, $\xi = \frac{1}{2}(\log 3 - 1)\sqrt{\log_2 x}$. We obtain $\nabla(n, \log 2) \geqslant 3$ with no more exceptions than given above; and this is sufficient.

Next, we determine to within a factor $(\log n)^{o(1)}$ the p.p. order of magnitude of the function $U(n,\alpha)$ defined in §4.2. It will be convenient to put

$$R_1(x,\xi) := \log_2 x \cdot \exp(-\xi^2/11), \tag{5.7}$$

$$R_2(x,\xi) := \exp(-\xi^2/50) + \xi^{-\beta}(\log_2 x)^{-\beta/2}(\log_3 x)^{4\beta}. \tag{5.8}$$

Theorem 53. *Let* $-\infty < \alpha \leqslant 1$. *Then*

$$U(n,\alpha) \sim 1 + (\log n)^{\log 3 - 1 + \alpha} \exp\{O(\sqrt{(\log_2 n \cdot \log_3 n)})\} \quad \text{p.p.} \tag{5.9}$$

More precisely, if $\xi = \xi(x) \to \infty$, $\xi \leqslant \sqrt{\log_2 x}$, *then we have*

$$\frac{\log U(n,\alpha)}{\log_2 x} \leqslant \left(\log 3 - 1 + \alpha + \frac{\xi(x)}{\sqrt{\log_2 x}}\right)^+ \tag{5.10}$$

for all except $\ll xR_1(x,\xi)$ *of the integers* $n \leqslant x$ *and*

$$\frac{\log U(n,\alpha)}{\log_2 x} \geqslant \left(\log 3 - 1 + \alpha - \frac{\xi(x)}{\sqrt{\log_2 x}}\right)^+ \tag{5.11}$$

for all except $\ll xR_2(x,\xi)$ *of the integers* $n \leqslant x$.

When $\alpha < 1 - \log 3$, the conclusion is that $U(n, \alpha) = 1$ p.p., that is the only pair of divisors counted is $d = d' = 1$. Since $U(n, \alpha)$ is an increasing function of α we may assume in the proof that $\alpha \geqslant -1$ (say). We put $u = \frac{3}{2}$, $v = x$, and notice that, for $-1 \leqslant \alpha \leqslant 1$, $\sqrt{x} < n \leqslant x$,

$$\nabla(n, \tfrac{1}{2}(\log x)^{\alpha}) \leqslant U(n, \alpha) \leqslant \nabla(n, 2(\log x)^{\alpha}).$$

The result follows from Theorems 50 and 51.

Remark. $R_1(x, \xi) \leqslant 1$ only when $\xi \geqslant \sqrt{(11 \log_3 x)}$, so that the upper bound (5.10) is useless unless the last term on the right is $\gg \sqrt{(\log_3 x / \log_2 x)}$. When $0 \leqslant \alpha \leqslant 1$ this is a weaker result than Theorem 45 where this term could be taken to be $c\sqrt{(\log_4 x / \log_2 x)}$. This is because the law of the iterated logarithm (Theorem 11) utilized in the proof of Theorem 45 is not available in the more general context of Theorem 50.

We now turn our attention to the function

$$E(n) := \min \{\log(d'/d) : d, d' | n, d < d'\}.$$

Theorem 54. *We have*

$$E(n) = (\log n)^{1 - \log 3} \exp \{O(\sqrt{(\log_2 n \cdot \log_3 n)})\} \quad \text{p.p.}; \tag{5.12}$$

more precisely, if $\psi(x) \to \infty$, $\psi(x) \leqslant \sqrt{\log_2 x}$, *then*

$$E(n) < (\log x)^{1 - \log 3} \exp \{\psi(x) \sqrt{\log_2 x}\} \tag{5.13}$$

for all except $\ll x R_2(x, \psi(x))$ *of the integers* $n \leqslant x$ *and*

$$E(n) > (\log x)^{1 - \log 3} \exp \{-\psi(x) \sqrt{\log_2 x}\} \tag{5.14}$$

for all except $\ll x R_1(x, \psi(x))$ *of the integers* $n \leqslant x$.

Proof. We use the fact that $E(n) > t$ if and only if $\nabla(n, t) = 1$. We apply Theorems 50, 51 with $u = \frac{3}{2}$, $v = x$ and $\xi = \psi(x) \mp 2/\sqrt{\log_2 x}$. It is then easy to check that the h in (5.2) satisfies

$$\tfrac{1}{3}(\log x)^{1 - \log 3} \leqslant h \leqslant \tfrac{1}{2}(\log x)^{1 - \log 3}.$$

The conclusion follows on setting $t = t_0 = \frac{1}{2} h \exp(-\xi\sqrt{w})$ in Theorem 50, and $t = t_1 = h \exp(\xi \sqrt{w})$ in Theorem 51. Indeed we obtain (in each case with the relevant number of exceptions) $\nabla(n, t_0) \leqslant \frac{3}{2}$ (whence $= 1$), $\nabla(n, t_1) > 1$.

In view of Theorems 53 and 54 it is tempting to conjecture that, if we define the functions $u(n, \alpha)$, $e(n)$ implicitly by the relations

$$U(n, \alpha) =: (\log n)^{\log 3 - 1 + \alpha} \exp \{u(n, \alpha) \sqrt{\log_2 n}\},$$

$$E(n) =: (\log n)^{1 - \log 3} \exp \{e(n) \sqrt{\log_2 n}\},$$

then $u(n, \alpha)$ and $e(n)$ both have distribution functions, but the present techniques to not seem delicate enough to decide this.

We close this section with a p.p. lower bound for the Δ-function. An upper bound follows in §5.4.

Theorem 55. *Let* $\gamma < \log 2/\log\{\log 3/(\log 3 - 1)\} = 0.287\,54\ldots$. *Then we have*

$$\Delta(n) > (\log_2 n)^\gamma \quad \text{p.p.}$$

Proof. Fix $\rho > \log 3/(\log 3 - 1)$ so that $\rho^\gamma < 2$. Set $J_1 := [\sqrt{\log_3 x}]$, $J_2 = [\log_3 x/\log \rho]$, and for each $j \in (J_1, J_2]$ put $u_j := \exp\exp \rho^j$ and $n_j = n(u_j, u_{j+1})$. Theorem 51, with $u = u_j$, $v = u_{j+1}$, $t = 1$ and $\xi = (\log 3 - \rho/(\rho - 1))\sqrt{(\rho^j(\rho - 1))}$ yields

$$\nabla(n_j, 1) > 1 \tag{5.15}$$

for all except $\ll_\rho x\rho^{-j\beta} j^{4\beta}$ of the integers $n \leqslant x$. Hence (5.15) holds simultaneously for all $j \in (J_1, J_2]$ and all but $o(x)$ integers $n \leqslant x$. In this circumstance, for every j there exist divisors d_{j0}, d_{j1} of n_j such that $0 < \log(d_{j1}/d_{j0}) \leqslant 1$, and all the products

$$\prod_{J_1 < j \leqslant J_2} d_{j\varepsilon(j)} \quad (\varepsilon(j) = 0 \text{ or } 1)$$

are divisors of n and fall in an interval of logarithmic length $J_2 - J_1$. Clearly this implies that

$$\Delta(n) \geqslant 2^{(J_2 - J_1)}/(J_2 - J_1)$$

and the result follows.

5.2 Proof of Theorem 50

Lemma 50.1. *Let* $\frac{3}{2} \leqslant u \leqslant x$, $\lambda \geqslant 1$, $\varepsilon \in (0, \frac{1}{2})$. *Then*

$$\max_{u \leqslant z \leqslant x} \left\{\Omega(n; u, z) - (1 + \varepsilon)\log\left(\frac{\log z}{\log u}\right)\right\} \leqslant \varepsilon\lambda$$

for all except $\ll x\varepsilon^{-2}(1 + \varepsilon)^{-\varepsilon\lambda}$ *of the integers* $n \leqslant x$.

Proof. This is an exercise on the parametric method of §0.5 and we just sketch the argument. Put $K = 1 + [\log(\log x/\log u)]$ and introduce the check-points

$$z_k = \exp(e^k \log u) \quad (0 \leqslant k \leqslant K).$$

The number of integers $n \leqslant x$ such that

$$\max_{0 \leqslant k \leqslant K} \{\Omega(n; u, z_k) - (1 + \varepsilon)(k - 1)\} > \varepsilon\lambda \tag{5.16}$$

does not exceed (for arbitrary $y \in [1, \frac{3}{2}]$),

$$\sum_{0 \leqslant k \leqslant K} \sum_{n \leqslant x} y^{\Omega(n; u, z_k) - (1+\varepsilon)(k-1) - \varepsilon\lambda}$$
$$\ll x \sum_{0 \leqslant k \leqslant K} \exp\{k(y - 1 - (1+\varepsilon)\log y) - \varepsilon\lambda \log y\}$$

and we put $y = 1 + \varepsilon$. The k-sum is $\ll \varepsilon^{-2}(1+\varepsilon)^{-\varepsilon\lambda}$. Now let $z \in [u, x]$. There exists a $k < K$ such that $z_k < z \leqslant z_{k+1}$. If n does not satisfy (5.16) then

$$\Omega(n; u, z) \leqslant \Omega(n; u, z_{k+1}) \leqslant (1+\varepsilon)k + \varepsilon\lambda$$
$$\leqslant (1+\varepsilon)\log\left(\frac{\log z}{\log u}\right) + \varepsilon\lambda$$

as required.

In the remainder of the chapter we use the notation

$$\sum_{n}^{u,v} \tag{5.17}$$

to mean that the sum is restricted to integers n for which $u < P^-(n)$ and $P^+(n) \leqslant v$. Note that $n = 1$ always satisfies these conditions if $\frac{3}{2} \leqslant u \leqslant v$.

Lemma 50.2. *Let $0 < y_0 < 2$. Then there exists a constant $c_0 = c_0(y_0) > 0$ such that, uniformly for $0 \leqslant y \leqslant y_0, \frac{3}{2} \leqslant u \leqslant \min(v, x)$,*

$$\sum_{n \leqslant x}^{u,v} y^{\Omega(n)} \ll_{y_0} x(\log x)^{y-1}(\log u)^{-y} \exp\left(-c_0 \frac{\log x}{\log v}\right).$$

Proof. We may assume $c_0 < \frac{1}{10}$ so that the sum over $n \leqslant \sqrt{x}$ may be easily seen to be negligible. In the remaining range $\sqrt{x} < n \leqslant x$ we use Rankin's method, in the form

$$y^{\Omega(n)} \leqslant x^{-\alpha/2} f_\alpha(n) \quad (\alpha > 0)$$

where f_α is the multiplicative function such that $f_\alpha(p^\nu) = (p^\alpha y)^\nu$. We put $\alpha = 2c_1/\log v$ where $c_1 = c_1(y_0)$ is such that $y_0 e^{2c_1} = 1 + \frac{1}{2}y_0 < 2$. Hence Theorem 01 applies to f_α and we get

$$\sum_{\sqrt{n} < n \leqslant x}^{u,v} y^{\Omega(n)} \leqslant x^{-\alpha/2} \sum_{n \leqslant x} f_\alpha(n)$$
$$\ll_{y_0} x^{1-\alpha/2}(\log x)^{-1} \exp\left\{\sum_{u < p \leqslant v} y p^{\alpha - 1}\right\}$$

which is within the asserted order of magnitude.

We may now embark on the proof of Theorem 50 which is similar to that of Theorem 45 except that the details are more cumbersome owing to the presence of the various parameters.

First, we note that we may suppose henceforth that $\xi \geqslant \xi_0$ where ξ_0 is any absolute constant, because the result is trivial otherwise. We may also without loss of generality restrict the parameter t to the range

$$\tfrac{1}{2}h\exp(-\xi\sqrt{w}) \leqslant t \leqslant \tfrac{1}{10}\log v. \tag{5.18}$$

For, if t is the lower limit here, (5.4) already implies $\nabla(n(u,v),t)=1$, and for smaller t we use the fact that ∇ decreases with t. If $t > 10^{-1}\log v$, we apply Lemma 50.1 with $\varepsilon = \xi/3\sqrt{w}(\leqslant \tfrac{1}{3})$ and $\lambda = w$, and we have

$$\nabla(n(u,v),t) \leqslant 3^{\Omega(n;u,v)} \leqslant 3^{w+(2\xi\sqrt{w})/3}$$

for all except $\ll x\varepsilon^{-2}(1+\varepsilon)^{-(\xi\sqrt{w})/3} \ll xw\exp(-\xi^2/11)$ integers $n \leqslant x$. This is sufficient because

$$3^{w+(2\xi\sqrt{w})/3} < \frac{10t}{\log v}\cdot 3^w\exp(\tfrac{3}{4}\xi\sqrt{w}) < h^{-1}t\exp(\xi\sqrt{w})$$

if ξ_0, and so ξ, is large enough.

Now we assume that (5.18) is realized and define

$$\nabla_1(n) := \sum_{d}^{u,v}\{1 : d|n, 1 < d \leqslant e^t\},$$

$$\nabla_2(n) := \sum_{d,d'}^{u,v}\{1 : dd'|n, 1 < d < d' \leqslant de^t\}.$$

Plainly, for all n, we have

$$\nabla(n(u,v),t) \leqslant 1 + 2(\nabla_1(n) + \nabla_2(n)) \tag{5.19}$$

and we begin by establishing an upper bound for $\nabla_1(n)$. We may suppose $e^t > u$, else $\nabla_1(n) = 0$. By Lemma 50.1 with $\varepsilon = \xi/3\sqrt{w}$, $\lambda = w$, we have in this circumstance

$$\Omega(n;u,e^t) \leqslant (1+\varepsilon)\log\left(\frac{t}{\log u}\right) + \varepsilon w < \frac{\log(h^{-1}t)}{\log 2} + \frac{1}{3}\xi\sqrt{w}$$

for all except $wx\exp(-\xi^2/11)$ of the integers $n \leqslant x$. Provided ξ_0 is large enough, we then have

$$\nabla_1(n) \leqslant 2^{\Omega(n;u,e^t)} < \tfrac{1}{4}h^{-1}t\exp(\xi\sqrt{w}). \tag{5.20}$$

We turn our attention to $\nabla_2(n)$. If the pair d, d' is to be counted, we must have

$$t \geqslant \log(d'/d) \geqslant \log(d'/(d'-1)) > 1/d' > e^{-t}d^{-1},$$

moreover $d > 1$, $P^-(d) > u$. Hence

$$d > \max(u, e^{-t}t^{-1}). \tag{5.21}$$

Define $I_1 := (u, \max(u,t^{-2})]$, $I_2 := (\max(u,t^{-2}), x^{1/4}]$ and $I_3 := (x^{1/4}, x^{1/2}]$ and for $j = 1,2,3$ let $\nabla_{2,j}(n)$ denote the contribution to $\nabla_2(n)$ from pairs $\{d,d'\}$ such that $d \in I_j$. I_1 is empty unless $u < t^{-2}$. In this case,

$$\sum_{n\leqslant x}\nabla_{21}(n) \leqslant \sum_{d\in I_1}^{u,v}\ \sum_{d<d'\leqslant de^t}^{u,v}\frac{x}{dd'}$$

and the sum over d' is $\leqslant d^{-1}(d(e^t-1)+1) \ll t + d^{-1} \ll t$ by (5.21), because $t \ll 1$. Also

$$\sum_{u<d\leqslant t^{-2}}^{u,v} \frac{1}{d} \leqslant \prod_{u<p\leqslant t^{-2}} \left(1-\frac{1}{p}\right)^{-1} \ll \frac{\log t^{-1}}{\log u} \ll \frac{w}{\log u}$$

(by (5.18)), and so

$$\sum_{n\leqslant x} \nabla_{21}(n) \ll xtw(\log u)^{-1} = xh^{-1}tw(3/e)^{-w}.$$

This implies that

$$\nabla_{21}(n) \leqslant h^{-1}t \tag{5.22}$$

for all but $\ll xw(3/e)^{-w} \ll x\exp(-\xi^2/11)$ of the integers $n \leqslant x$.

The quantities $\nabla_{22}(n)$ and $\nabla_{23}(n)$ will be dealt with by the parametric method of §0.5; we begin by applying Lemma 50.1 with $\varepsilon = \xi/3\sqrt{w}$, $\lambda = w$, to obtain, with $\ll xw\exp(-\xi^2/11)$ exceptional $n \leqslant x$,

$$\Omega(n; u, d') \leqslant (1+\varepsilon)\log\left(\frac{\log d'}{\log u}\right) + \varepsilon w \tag{5.23}$$

simultaneously for all d', $d' | n(u,v)$. For the integers n for which (5.23) holds, we have

$$\nabla_{2,j}(n) \leqslant y^{-\varepsilon w}\nabla^*_{2,j}(n) \quad (j=2,3) \tag{5.24}$$

with

$$\nabla^*_{2,j}(n) := \sum_{\substack{d|n \\ d\in I_j}}^{u,v} \sum_{\substack{d'|n/d \\ d<d'\leqslant de^t}}^{u,v} y^{\Omega(n;u,d')}\left(\frac{\log d'}{\log u}\right)^{-(1+\varepsilon)\log y} \tag{5.25}$$

where $y \in (0,1]$ is a free parameter to be chosen later. We are going to show that, for suitable y, $\nabla^*_{2,j}(n)$ is small on average. We begin with $j=2$. We have

$$S_{22} := \sum_{n\leqslant x} \nabla^*_{22}(n) \leqslant \sum_{d\in I_2}^{u,v} \sum_{d<d'\leqslant de^t}^{u,v} y^{\Omega(dd')} \times \left(\frac{\log d'}{\log u}\right)^{-(1+\varepsilon)\log y} \sum_{m\leqslant x/dd'} y^{\Omega(m;u,d')}.$$

Now $dd' \leqslant d^2e^t \leqslant x^{1/2}v^{1/10} \leqslant x^{3/5}$ by (5.18), whence the innermost sum may be estimated by Theorem 01 and we obtain

$$S_{22} \ll x \sum_{d\in I_2}^{u,v} \frac{y^{\Omega(d)}}{d} \sum_{d<d'\leqslant de^t}^{u,v} \frac{y^{\Omega(d')}}{d'} \left(\frac{\log d'}{\log u}\right)^A$$

where $A = y - 1 - (1+\varepsilon)\log y$. When $t > 1$ the inner sum may be majorized by partial summation, using the estimate

$$\sum_{d'\leqslant D}^{u,v} y^{\Omega(d')} \ll D(\log D)^{y-1}(\log u)^{-y} \quad (D > u)$$

which follows from Theorem 01 (dropping the condition involving v). We obtain that the sum over d' above is

$$\ll \int_a^{a+b} z^B \, \mathrm{d}z$$

with $a = \log d/\log u$, $b = t/\log u$, $B = 2y - 2 - (1+\varepsilon)\log y$. If we assume, as will be the case, that $B \leqslant 0$, this is

$$\ll ba^B = \frac{t}{\log u}\left(\frac{\log d}{\log u}\right)^B. \tag{5.26}$$

From Theorem 03 (Shiu), this bound is equally valid when $t \leqslant 1$ provided $d > t^{-2}$. Hence the d'-sum is majorized by (5.26) for all $d \in I_2$ and we infer that

$$S_{22} \ll \frac{xt}{\log u} \sum_{d \in I_2}^{u,v} \frac{y^{\Omega(d)}}{d}\left(\frac{\log d}{\log u}\right)^B \ll \frac{xt}{\log u} \frac{\exp\{(B+y)w\}}{(B+y)}$$

by partial summation, appealing to Lemma 50.2. We select $y = (1+\varepsilon)/3$, so that $B \leqslant \frac{4}{3}\log\frac{9}{4} - \frac{10}{9} < 0$ and $B + y = \log 3 - 1 + \varepsilon \log 3 - Q(1+\varepsilon) \geqslant \log 3 - 1$. Hence

$$S_{22} \ll xh^{-1} t \exp\left(\frac{\log 3}{3}\xi\sqrt{w}\right)$$

and by (5.24) we obtain that

$$\nabla_{22}(n) \leqslant \tfrac{1}{12} h^{-1} t \exp(\xi\sqrt{w}) \tag{5.27}$$

for all but $\ll x\exp\{-(1-\frac{2}{3}\log 3)\xi\sqrt{w}\} \ll xw\exp(-\xi^2/11)$ of the integers $n \leqslant x$ satisfying (5.23).

The case of $\nabla_{23}^*(n)$ is very similar and we only give the main steps. By Theorem 01,

$$S_{23} := \sum_{n \leqslant x} \nabla_{23}^*(n) \ll x\left(\frac{\log x}{\log u}\right)^{-(1+\varepsilon)\log y}$$

$$\times \sum_{d \in I_3}^{u,v} \frac{y^{\Omega(d)}}{d} \sum_{\substack{d < d' \\ \leqslant \min(de^t, x/d)}}^{u,v} \frac{y^{\Omega(d')}}{d'}\left(\frac{\log(2x/dd')}{\log u}\right)^{y-1}.$$

We drop the condition $P^+(d') \leqslant v$ in the inner sum and, if $t \geqslant \frac{1}{2}$, we estimate by Theorem 01 again, and partial integration. We obtain

$$\ll \begin{cases} (\log d)^{y-1}\left(\log\dfrac{2x}{d^2 e^t}\right)^{y-1} t(\log u)^{1-2y} & (de^t \leqslant \sqrt{x}), \\[2ex] y^{-1}(\log d)^{y-1}\left(\log\dfrac{2x}{d^2}\right)^{y}(\log u)^{1-2y} & (de^t > \sqrt{x}). \end{cases}$$

Now, when $de^t \leqslant \sqrt{x}$, we have $(\log(2x/d^2e^t))^{y-1} \ll (\log 2x/d^2)^{y-1}$ and so for every d the d'-sum is

$$\ll y^{-1}(\log x)^{y-1}(\log(2x/d^2))^{y-1}t(\log u)^{1-2y}. \tag{5.28}$$

If $t < \frac{1}{2}$, we apply Theorem 03 (Shiu) which is applicable because $d > t^{-2}$. We find that the upper bound (5.28) is still valid. Whence

$$S_{23} \ll xy^{-1}t(\log x)^D(\log u)^{-y-D} \sum_{d \in I_3}^{u,v} \frac{y^{\Omega(d)}}{d}\left(\log\frac{2x}{d^2}\right)^{y-1}$$

where $D := y - 1 - (1+\varepsilon)\log y$. By Lemma 50.2, and partial summation, the sum over d is

$$\ll y^{-1}(\log x)^{2y-1}(\log u)^{-y}\exp\left\{-\frac{c_0(1)}{4}\frac{\log x}{\log v}\right\}$$

and so

$$S_{23} \ll xy^{-2}\left(\frac{\log x}{\log u}\right)^E \frac{t}{\log u}\exp\left\{-\frac{c_0(1)}{4}\frac{\log x}{\log v}\right\},$$

with $E = 3y - 2 - (1+\varepsilon)\log y$. We select $y = (1+\varepsilon)/3$ as before and obtain

$$S_{23} \ll xh^{-1}t\exp\left(\frac{\log 3}{3}\xi\sqrt{w}\right).$$

Thus, the upper bound (5.27) is equally valid for $\nabla_{23}(n)$, with the same estimate for the number of exceptional integers $n \leqslant x$. Together with (5.22) and (5.19), this completes the proof of Theorem 50.

5.3 Proof of Theorem 51

Our method is based on an induction process. The basic idea is to consider the product of the first k prime factors of n and try to keep as close control as possible over what happens when we multiply by the $(k+1)$st prime factor. However the function $n \to \prod\{p_j(n): j \leqslant k\}$ has the drawback of not being multiplicative, and we introduce the slightly different function n_k (which also takes account of u) to ease the technical difficulties.

The parameters u, v, w, h, ξ, x being fixed as in the statement of the theorem, we put $L = [w - \frac{2}{5}\xi\sqrt{w}]$, $M := [w - \frac{1}{5}\xi\sqrt{w}]$, and define

$$u_k := \exp(e^k \log u) \ (k = 0, 1, 2, \ldots)$$

so that $u \leqslant u_k \leqslant v$ for all $k \leqslant M$; moreover

$$\exp(-\tfrac{2}{5}\xi\sqrt{w} - 1) < \log u_k/\log v \leqslant \exp(-\tfrac{1}{5}\xi\sqrt{w})$$

for $L \leqslant k \leqslant M$. Now we define, for every n,

$$n_k := \prod_{\substack{p \mid n \\ u < p \leqslant u_k}} p. \tag{5.29}$$

This is a multiplicative function of n and when $k \leqslant M$ it is a squarefree divisor of $n(u, v)$. The normal order of $\omega(n_k)$ is k (if $k \to \infty$) and so n_k mimics the product of the first k prime factors of $n(u, v)$.

Lemma 51.1. *For each fixed k, $L \leqslant k \leqslant M$, we have*

$$k - \tfrac{1}{5}\xi\sqrt{k} \leqslant \omega(n_k) \leqslant 2k \tag{5.30}$$

for all but $\ll x\exp(-\xi^2/50)$ of the integers $n \leqslant x$.

This is a straightforward application of the parametric method of §0.5 and we omit the proof.

Lemma 51.2. *Let $1 < T \leqslant w$, $0 < \alpha < 1$. Then we have*

$$\min_{T \leqslant m \leqslant k} m^{-1}(\omega(n_k) - \omega(n_{k-m})) > \alpha \tag{5.31}$$

for all except $\ll xQ(\alpha)^{-1}\exp(-TQ(\alpha))$ of the integers $n \leqslant x$ (with $Q(\alpha) = \alpha\log\alpha - \alpha + 1$ as in (0.22)).

Proof. The number of exceptional n does not exceed

$$\sum_{T \leqslant m \leqslant k}\sum_{n \leqslant x} \alpha^{\omega(n_k) - \omega(n_{k-m}) - m\alpha}$$
$$\ll \sum_{T \leqslant m \leqslant k} xe^{-mQ(\alpha)} \ll xQ(\alpha)^{-1}\exp(-TQ(\alpha))$$

using Theorem 01.

Our next lemma concerns the function $\rho(n, \theta)$ defined in (3.17). Recall that, for squarefree n,

$$\rho(n, \theta) = \prod_{p \mid n}(1 + 2\cos(\theta\log p)).$$

We also use the notation

$$\omega_\theta(n) := \omega(n, \exp(1/\theta)) \quad (\theta > 0).$$

Lemma 51.3. *Let $\theta > 0$, $L \leqslant k \leqslant M$. Then we have*

$$\sum_{n \leqslant x} 3^{-\omega(n_k) - \omega_\theta(n_k)}\rho(n_k, \theta)^2 \ll x(\log(3 + \theta))^4. \tag{5.32}$$

Proof. We distinguish several cases, according to the value of θ. When $\theta \leqslant (\log u_k)^{-1}$, we have $\omega_\theta(n_k) = \omega(n_k)$ for all n and the summand in (5.32) is $\leqslant 1$. The conclusion is trivial. When $(\log u_k)^{-1} < \theta \leqslant (\log u)^{-1}$, we apply Theorem 01 to the left-hand side of (5.32) which yields the bound

$$\ll x\exp\left\{\sum_{u < p \leqslant \exp(1/\theta)} \frac{1}{p}(-1 + \rho(p, \theta)^2/9) + \sum_{\exp(1/\theta) < p \leqslant u_k} \frac{1}{p}(-1 + \rho(p, \theta)^2/3)\right\}.$$

The first of the p-sums is $\leqslant 0$. The second falls within the range of applicability of Lemma 30.1 with $f(v) = -1 + (1 + 2\cos v)^2/3$. Since f has mean value zero over $[0, 2\pi)$ we find that the sum is $O(1)$. This implies the asserted result.

When $\theta > (\log u)^{-1}$ we have $\omega_\theta(n_k) \equiv 0$ and Theorem 01 provides the bound

$$\ll x \exp\left\{ \sum_{u < p \leqslant u_k} \frac{1}{p}(-1 + \rho(p, \theta)^2/3) \right\}.$$

If $\theta \leqslant \exp\sqrt{\log u}$, we see again that the sum over p is $O(1)$ by Lemma 30.1. If $\theta > \exp\sqrt{\log u}$, we apply the lemma to the sub-sum corresponding to the range $\exp\{(\log(3+\theta))^2\} < p \leqslant u_k$ – this is $O(1)$ – and the complementary sub-sum is majorized trivially, just writing $\rho(p, \theta)^2 \leqslant 9$. It does not exceed $4\log_2(3+\theta) + O(1)$, which gives the result stated.

Our next lemma is the key to the proof. Here β is the small positive number defined by (5.5). Furthermore, we set

$$H := h\exp(\tfrac{1}{2}\xi\sqrt{w}). \tag{5.33}$$

Lemma 51.4 (Fundamental Lemma). *Let $L \leqslant k \leqslant M$, $0 < \varepsilon < 1/100$. Then we have*

$$\int_{-1/H}^{1/H} \rho(n_k, \theta)^2 \, d\theta \leqslant 3^{2\omega(n_k)}(\varepsilon \log u_k)^{-1} \tag{5.34}$$

for all except $\ll x(\varepsilon^\beta + \exp(-\xi^2/50))$ of the integers $n \leqslant x$.

Proof. Set $\mathscr{D}_1 := [0, (6\varepsilon\log u_k)^{-1}]$, $\mathscr{D}_2 := ((6\varepsilon\log u_k)^{-1}, (\log u)^{-1}]$, $\mathscr{D}_3 := ((\log u)^{-1}, H^{-1}]$, and define

$$A_j(n_k) := \int_{\mathscr{D}_j} 3^{-2\omega(n_k)}\rho(n_k, \theta)^2 \, d\theta \quad (1 \leqslant j \leqslant 3). \tag{5.35}$$

Our conclusion will follow if we can show that

$$A_j(n_k) \leqslant (6\varepsilon\log u_k)^{-1} \quad (1 \leqslant j \leqslant 3) \tag{5.36}$$

for non-exceptional integers; and since the integrand in (5.35) is $\leqslant 1$, this is certainly true for $j = 1$.

Let $\delta = \delta(\varepsilon)$, $1 < \delta < \log 3$, be a parameter to be chosen later. We apply Lemma 51.2 with $T = [\log(1/6\varepsilon)]$ and $\alpha = \delta/\log 3$. Since $m := [\log(\theta\log u_k)] \geqslant T$ for $\theta \in \mathscr{D}_2$, we obtain that

$$\omega(n_k) - \omega_\theta(n_k) \geqslant \omega(n_k) - \omega(n_{k-m}) > \alpha[\log(\theta\log u_k)]$$

holds simultaneously for all $\theta \in \mathscr{D}_2$ and all $n \leqslant x$ except those which belong to a fixed set (independent of θ) of cardinality

$$\ll xQ(\alpha)^{-1}\varepsilon^{Q(\alpha)}. \tag{5.37}$$

For the remaining integers, we have

$$3^{-\omega(n_k)} < 3^{1-\omega_\theta(n_k)}(\theta \log u_k)^{-\delta}$$

whence $A_2(n_k) < 3A_2^*(n_k)$ where

$$A_2^*(n_k) := \int_{\mathscr{D}_2} 3^{-\omega(n_k)-\omega_\theta(n_k)} \rho(n_k, \theta)^2 \frac{d\theta}{(\theta \log u_k)^\delta}.$$

Now Lemma 51.3 implies (because $\theta = O(1)$),

$$\sum_{n \leqslant x} A_2^*(n_k) \ll x \int_{\mathscr{D}_2} (\theta \log u_k)^{-\delta} \, d\theta \ll x(\delta - 1)^{-1} \varepsilon^{\delta - 1} (\log u_k)^{-1}$$

whence $A_2^*(n_k) \leqslant (18\varepsilon \log u_k)^{-1}$ for all except $\ll x(\delta - 1)^{-1}\varepsilon^\delta$ of the integers $n \leqslant x$, and we deduce that (5.36) holds for $j = 2$ and all but

$$\ll x\{(\delta - 1)^{-1}\varepsilon^\delta + Q(\alpha)^{-1}\varepsilon^{Q(\alpha)}\} \tag{5.38}$$

integers $n \leqslant x$.

Next we consider $A_3(n_k)$. From Lemma 51.1 we have

$$A_3(n_k) \leqslant 3^{-k + (\xi\sqrt{k})/5} A_3^*(n_k) \tag{5.39}$$

with

$$A_3^*(n_k) := \int_{\mathscr{S}_3} 3^{-\omega(n_k)} \rho(n_k, \theta)^2 \, d\theta$$

for all except $\ll x \exp(-\xi^2/50)$ of the integers $n \leqslant x$. Noticing that $\omega_\theta(n_k) \equiv 0$ for $\theta \in \mathscr{D}_3$, we may infer from Lemma 51.3 that

$$\sum_{n \leqslant x} A_3^*(n_k) \ll x \int_{\mathscr{D}_3} (\log(3 + \theta))^4 \, d\theta \ll x H^{-1} w^4. \tag{5.40}$$

It may be checked that, for $L \leqslant k \leqslant M$, we have

$$H^{-1} w^4 \cdot 3^{-k + (\xi\sqrt{k})/5} \leqslant 3w^4 (\log u_k)^{-1} \exp\left\{-\left(\frac{7 - 6\log 3}{10}\right)\xi\sqrt{w}\right\}$$

so that, assuming as we may that $\xi \geqslant 1$, we may deduce from (5.39) and (5.40) that (5.36) holds for $j = 3$ with $\ll x(\varepsilon + \exp(-\xi^2/50))$ exceptions among the integers $n \leqslant x$.

We put $\delta = 1 + \frac{1}{2}\sqrt{\varepsilon}$ $(< \log 3)$ and we see that (5.38) is within the order of magnitude asserted. Collecting the various estimates together we obtain the result stated.

We are now in a position to embark on the proof of the second main theorem.

Proof of Theorem 51. We assume that $\xi \geqslant \xi_0$ where ξ_0 is a sufficiently large absolute constant. We shall establish that the slightly better bound

$$\nabla(n(u, v), t) > H^{-1} t \tag{5.41}$$

(where H is as in (5.33)) holds with the number of exceptions claimed, in the narrower range

$$H \leqslant t \leqslant \log v \cdot \exp(-\tfrac{1}{2}\xi\sqrt{w}). \tag{5.42}$$

This is sufficient to imply (5.6) for $0 \leqslant t \leqslant \log v$. For $t < H$, (5.6) is weaker than the trivial lower bound $\nabla \geqslant 1$. If t_1 denotes the upper limit in (5.42) then, for $t_1 < t \leqslant \log v$, we shall have $\nabla(n(u,v),t) \geqslant \nabla(n(u,v),t_1) > H^{-1}t_1 \geqslant h^{-1}t\exp(-\xi\sqrt{w})$ as required. It remains to obtain (5.41).

With the notation (5.17) as before we introduce the function

$$V(n,z) := \sum_{d,d'}^{u,v} \{1 : d, d'|n, (d,d') = 1, |\log(d'/d) - z| \leqslant t\}$$

so that $\nabla(n(u,v),t) = V(n,0)$. For $|z| > t + \log n$, evidently $V(n,z) = 0$. The idea is as follows: suppose we know that, for some particular z_0, $V(n,z_0)$ is large. Then, if $p \nmid n$, $\log p \approx z_0$, $V(pn,0)$ should be large. With this in mind we aim to show that, for any k, $L \leqslant k \leqslant M$, $V(n_k, z)$ is so often large that, for small r, $V(n_{k+r}, 0)$ cannot be small too frequently. The argument can be viewed in terms of conditional probabilities, and concludes with an iterative procedure.

Thus we introduce the set of points

$$\mathscr{L}(n) := \{z \in \mathbb{R} : V(n,z) > H^{-1}t\}$$

where $V(n,z)$ is large. The Lebesgue measure of $\mathscr{L}(n)$ is denoted by $\lambda(n)$. Plainly $\lambda(n_k) \leqslant 2(t + \log n_k)$ and we begin by showing that this easy upper bound, if it be multiplied by ε, persists as a lower bound valid for most of the integers $n \leqslant x$. Let m be squarefree, $u < P^-(m) \leqslant P^+(m) \leqslant v$. Then

$$3^{\omega(m)}t \leqslant H^{-1}t(t + \log m) + \left(2\pi\lambda(m)t^2 \int_{-1/t}^{1/t} \rho(m,\theta)^2 \, d\theta\right)^{1/2}. \tag{5.43}$$

To see this, observe that

$$2 \cdot 3^{\omega(m)}t = \int_{-\infty}^{\infty} V(m,z) \, dz$$

and majorize the integrand by $H^{-1}t$ for $z \notin \mathscr{L}(m)$. This yields

$$2 \cdot 3^{\omega(m)}t \leqslant 2H^{-1}t(t + \log m) + \int_{\mathscr{L}(m)} V(m,z) \, dz. \tag{5.44}$$

Now

$$\begin{aligned} V(m,z) &\leqslant 2 \sum_{\substack{dd'|m \\ (d,d')=1}} \left(\frac{\sin((\log(d'/d) - z)/2t)}{(\log(d'/d) - z)/2t}\right)^2 \\ &= 2t \int_{-1/t}^{1/t} e^{i\theta z}(1 - |\theta t|)\rho(m,\theta) \, d\theta \end{aligned} \tag{5.45}$$

and Parseval's formula yields

$$\int_{-\infty}^{\infty} V(m,z)^2 \, dz \leqslant 8\pi t^2 \int_{-1/t}^{1/t} (1 - |\theta t|)^2 \rho(m,\theta)^2 \, d\theta.$$

We apply Cauchy's inequality to the integral in (5.44) and obtain (5.43).

Let $\varepsilon = \varepsilon(\xi, w) > w^{-1}$ be a small positive number to be chosen later and put $\eta = \varepsilon^{\beta} + \exp(-\xi^2/50)$ with β as in (5.5). For any $k \in [L, M]$ we have by Lemma 51.4

$$\int_{-1/H}^{1/H} \rho(n_k, \theta)^2 \, d\theta \leqslant 3^{2\omega(n_k)}(\varepsilon \log u_k)^{-1} \tag{5.46}$$

for all except $\ll \eta x$ of the integers $n \leqslant x$. Let us prove that for t satisfying (5.42) we have, also with $\ll \eta x$ exceptions

$$H^{-1}t(t + \log n_k) \leqslant \tfrac{1}{2} \cdot 3^{\omega(n_k)}t. \tag{5.47}$$

To see this, first note that by Theorem 07 there exists an absolute constant c_0 such that

$$\log n_k \leqslant c_0 \log w \cdot \log u_k \tag{5.48}$$

for all but $\ll \eta x$ of the integers $n \leqslant x$. Next, observe that, for $k \geqslant L \geqslant w - \frac{2}{5}\xi\sqrt{w} - 1$, Lemma 51.1 implies

$$w \leqslant \omega(n_k) + \tfrac{3}{5}\xi\sqrt{w} + 1$$

for $(1 + O(\eta))x$ integers $n \leqslant x$. When this inequality and (5.48) hold, we have

$$\begin{aligned} H^{-1}\log n_k &= 3^w (\log v)^{-1} \log n_k \cdot \exp(-\tfrac{1}{2}\xi\sqrt{w}) \\ &\leqslant 3c_0 \log w \cdot \exp\left\{-\left(\frac{7 - 6\log 3}{10}\right)\xi\sqrt{w}\right\} \cdot 3^{\omega(n_k)} \\ &\leqslant \tfrac{1}{4} \cdot 3^{\omega(n_k)} \end{aligned}$$

provided ξ_0, and hence ξ, is sufficiently large: we have used the inequality immediately following (5.40). Moreover, by (5.42) and the definition of H,

$$\begin{aligned} H^{-1}t &\leqslant 3^w \exp(-\xi\sqrt{w}) \leqslant 3^{\omega(n_k)+1} \exp\left\{-\left(1 - \frac{3\log 3}{5}\right)\xi\sqrt{w}\right\} \\ &\leqslant \tfrac{1}{4} \cdot 3^{\omega(n_k)} \end{aligned}$$

and (5.47) follows. Substituting $m = n_k$ in (5.43) and taking (5.46) and (5.47) into account, we finally obtain the desired lower bound for $\lambda(n_k)$. For t as in (5.42) and $k \in [L, M]$, we have

$$\lambda(n_k) \geqslant \frac{\varepsilon}{8\pi} \log u_k \tag{5.49}$$

for all except $\ll \eta x$ of the integers $n \leqslant x$.

We may now take up the final part of the proof. We proceed by induction on k, our aim being to establish a non-trivial upper bound for

$$N_k := \operatorname{card}\{n : n \leqslant x, \nabla(n_k, t) \leqslant H^{-1}t\}.$$

It will be useful to bear in mind that this is a decreasing function of k. We introduce the number N_k^* of those n counted by N_k which satisfy, in addition, the three conditions

(a) $\log n_k \leqslant R \log u_k$,

(b) $k - \xi\sqrt{k} \leqslant \omega(n_k) \leqslant 2k$,

(c) $\lambda(n_k) \geqslant \varepsilon \log u_k$,

where we have set $R := c_0 \log w$. From (5.48), Lemma 51.1, and (5.49) we deduce that

$$N_k \leqslant N_k^* + c_1 \eta x \tag{5.50}$$

where c_1 is absolute. Let D_k denote the number of those n counted by N_k^* which have two prime factors p, q such that

(d) $2R \log u_k < \log p \leqslant 3R \log u_k$,

(e) $\log p < \log q \leqslant \log p + R \log u_k$,

(f) $\log q - \log p \in \mathscr{L}(n_k)$.

Condition (f) implies $V(n_k, \log q - \log p) > H^{-1}t$ so that the inequality $|\log(pd'/qd)| \leqslant t$ has more than $H^{-1}t$ solutions in pairs $\{d, d'\}$ such that $dd' | n_k$. Put $r = 1 + [\log 4R]$. By (d) and (e), we have $p < q \leqslant u_{k+r}$ and it follows that if n is counted by D_k, it cannot be counted by N_{k+r}. Therefore

$$D_k \leqslant N_k - N_{k+r}. \tag{5.51}$$

Denote generically by m^* an integer equal to n_k for some n counted by N_k^* and by b an integer for which $P^-(b) > \exp(4R \log u_k)$. Then D_k counts at least all the integers of the form $n = am^*dpqb$, with $a \leqslant u$, $d | m^*$, p, q as in (d)–(f) (with $\mathscr{L}(m^*)$ in (f)). Hence

$$D_k \geqslant \sum_{a \leqslant u} \sum_{m^*} \sum_{d | m^*} \sum_{p,q} \sum_{b \leqslant x/am^*dpq} 1.$$

The innermost sum may be estimated by Theorem 06, on noticing that $adm^*pq \leqslant u(m^*)^2 pq \leqslant \exp(10R \log u_k)$ (by (a), (d), (e) above), whence, provided ξ_0 is sufficiently large, $\exp(4R \log u_k) < \frac{1}{2}x/am^*dpq$. We obtain

$$D_k \gg xe^{-k} R^{-1} \sum_{m^*} \frac{1}{\varphi(m^*)} \sum_{p,q} \frac{1}{pq}. \tag{5.52}$$

When m^* and p are fixed, q covers a union of at most $3^{\omega(m^*)} \leqslant 9^k$ intervals with total logarithmic length at least equal to the Lebesgue measure of $\mathscr{L}(m^*) \cap [0, \infty) = \lambda(m^*)/2$ and with end-points having logarithms $\asymp R \log u_k$. By (c) above, and the Prime Number Theorem, we have

$$\sum \frac{1}{q} \gg (R \log u_k)^{-1} \sum \frac{\log q}{q}$$
$$\gg \lambda(m^*)(R \log u_k)^{-1} + O(9^k \exp(-\sqrt{(R \log u_k)})).$$

We shall choose $\varepsilon > w^{-1} \geqslant (2k)^{-1}$, and the error term above is therefore negligible – the q-sum is $\gg \varepsilon R^{-1}$. The p-sum is then $\gg 1$ and

$$D_k \gg xe^{-k}\varepsilon R^{-2}\sum_{m^*}\varphi(m^*)^{-1}. \tag{5.53}$$

Now we also have

$$N_k^* \leqslant \sum_{m^*}\sum_{d\leqslant x/m^*}{}' 1$$

where the dash denotes that summation is restricted to those d for which $p|d$ implies either $p|m^*$ or $p\notin(u, u_k]$ – writing the integers n counted by N_k^* in the form $n = m^*d$, where $m^* = n_k$. Since $\log m^* \leqslant R\log u_k$ this inner sum can be estimated by Theorem 01 and we obtain

$$N_k^* \ll xe^{-k}\sum_{m^*}\varphi(m^*)^{-1}$$

which, compared with (5.53), yields

$$D_k \geqslant c_2\varepsilon R^{-2}N_k^*. \tag{5.54}$$

Suppose now that $N_M \geqslant 2c_1\eta x$ where c_1 is the constant appearing in (5.50). Then we have, for all $k\in[L, M]$,

$$N_k^* \geqslant N_k - c_1\eta x \geqslant N_k - \tfrac{1}{2}N_M \geqslant \tfrac{1}{2}N_k,$$

whence $D_k \geqslant \frac{1}{2}c_2\varepsilon R^{-2}N_k$ by (5.54) and $N_{k+r} \leqslant N_k(1 - \frac{1}{2}c_2\varepsilon R^{-2})$ by (5.51). By iteration,

$$N_M \leqslant N_L(1 - \tfrac{1}{2}c_2\varepsilon R^{-2})^{[(M-L)/r]} \leqslant x\exp\left(-c_3\frac{\varepsilon\xi\sqrt{w}}{R^2\log R}\right),$$

and so, in any event,

$$N_M \leqslant x\left\{2c_1\eta + \exp\left(-c_3\frac{\varepsilon\xi\sqrt{w}}{R^2\log R}\right)\right\}.$$

We select $\varepsilon = \xi^{-1}w^{-1/2}(\log w)^4 > w^{-1}$ which yields

$$N_M \ll x(\varepsilon^\beta + \exp(-\xi^2/50))$$

and achieves the conclusion of Theorem 51.

5.4 A p.p. upper bound for the Δ-function

We have seen that the Δ-function plays a central rôle in the theory of divisors and the question of its 'normal order' is naturally of interest. As a complement to Theorem 55 we prove the following result.

Theorem 56. *Let* $\psi(n)\to\infty$ *as* $n\to\infty$. *Then*

$$\Delta(n) < \psi(n)\log_2 n \quad \text{p.p.} \tag{5.55}$$

In the next chapter (Theorem 60) we prove that the average order of $\Delta(n)$

is at least $c \log_2 n$. We believe that this mean is dominated by those integers n such that $\omega(n) \sim \log_2 n$ and make the following

Conjecture. $\Delta(n)/\log_2 n$ *has a continuous distribution function.*

Notice that, if this were true, Theorem 56 would be optimal and Theorem 55 could be sharpened to $\Delta(n) > \psi(n)^{-1} \log_2 n$ p.p.

The proof of Theorem 56 is based on a study of the high moments of the function

$$\Delta(n; u) = \operatorname{card}\{d: d|n, \mathrm{e}^u < d \leqslant \mathrm{e}^{u+1}\}.$$

Set

$$M_q(n) := \int_{-\infty}^{\infty} \Delta(n; u)^q \, \mathrm{d}u \quad (n, q \in \mathbb{Z}^+).$$

It is plain that, for fixed n and $q \to \infty$, $M_q(n)^{1/q} \to \Delta(n)$. Moreover we have (Theorem 72) that

$$\Delta(n) \leqslant 2M_q(n)^{1/q}. \tag{5.56}$$

We use an induction analogous to that involved in the proof of Theorem 51 but we must now consider the occurrence of the prime factors one at a time, separately. For technical reasons we rule out the very small, and very large, prime factors from this process. Thus we fix a function $L = L(x)$ such that $L(x) \to \infty$, $L(x) < \sqrt{\log_2 x}$, and define, for $n \leqslant x$,

$$K = K(n) := \max\{k: 1 \leqslant k \leqslant \omega(n), \log p_k(n) \leqslant \mathrm{e}^{-L} \log x\}$$

and

$$n_k := \begin{cases} \prod_{L < j \leqslant k} p_j(n) & (L < k \leqslant K(n)), \\ n_K & (k > K(n)). \end{cases}$$

We may assume that L is integer valued, so that, for each n, $\omega(n_k) = \min(k, K) - L$ $(k, K > L)$. From Theorem 10 we also have

$$L < K(n) < M := [2 \log_2 x] \quad \text{p.p.x.} \tag{5.57}$$

We shall need an improvement, p.p.x, on the factor 2 on the right of (5.56) when $n = n_k$. This is provided by the following result which is an easy application of Theorem 50.

Lemma 56.1. *Let* $\varepsilon \in (0, 1)$. *Then we have*

$$\max_{\substack{L < k \leqslant M \\ q \geqslant 1}} \{\Delta(n_k) - (3/\mathrm{e})^{(1+\varepsilon)k/q} M_q(n_k)^{1/q}\} \leqslant 1 \quad \text{p.p.x.}$$

Proof. Let k, q be fixed, and u_0 chosen so that $\Delta(n_k; u_0) = \Delta(n_k)$. Let d be the smallest divisor of n_k in $(\mathrm{e}^{u_0}, \mathrm{e}^{u_0+1}]$. We may assume $d \neq n_k$, for we should then have $\Delta(n_k) = 1$ and could set $u_0 = -1$, $d = 1$. Let d' be the smallest divisor of n_k greater than d. By definition of the function E (see §5.1) we have

$$\log d' - \log d \geqslant E(n_k)$$

and, since $\Delta(n_k; u) \geqslant \Delta(n_k) - 1$ for $\log d \leqslant u < \log d'$,

$$M_q(n_k) \geqslant \int_{\log d}^{\log d'} (\Delta(n_k) - 1)^q \, du \geqslant (\Delta(n_k) - 1)^q E(n_k).$$

Therefore the desired result will follow from

$$\min_{L<k\leqslant M} (3/e)^{(1+\varepsilon)k} E(n_k) \geqslant 1 \quad \text{p.p.}x. \tag{5.58}$$

To establish this, observe that, by the corollary to Theorem 10, there exists a subset S of the integers $n \leqslant x$, with cardinality $(1 + o(1))x$, on which

$$\max_{L<k\leqslant M} \{\log_2 P^+(n_k) - (1 + \varepsilon/2)k\} \leqslant 0.$$

For $n \in S$, we have, in the notation of §5.1, that $n_k | n(u, v)$ $(u = \frac{3}{2},\ v = \min(x, \exp\exp((1 + \varepsilon/2)k)))$ for every $k \in (L, M)$. We apply Theorem 50 with this u and v, and $\xi = \frac{1}{25}\varepsilon\sqrt{k}$. We obtain

$$\nabla(n_k, (3/e)^{-(1+\varepsilon)k}) = 1 \quad (L < k \leqslant M) \tag{5.59}$$

for all but $\ll x\exp(-\varepsilon^2 k/7000)$ members of S: that is, (5.59) holds, simultaneously for all $k \in (L, M]$, p.p.x. This implies (5.58) and concludes the proof of Lemma 56.1.

Proof of Theorem 56. We are going to prove that

$$\Delta(n_k) \ll K(n) \quad \text{p.p.}x. \tag{5.60}$$

It follows from Lemma 50.1 that $\Omega(n/n_K) < (2 + o(1))L$ p.p.x; moreover from one of the elementary properties of Δ (see Lemma 61.1) we also have $\Delta(n) \leqslant \tau(n/n_K)\Delta(n_K)$, whence

$$\Delta(n) < 5^L \Delta(n_K) \quad \text{p.p.}x.$$

Together with (5.60), this implies (5.55).

The starting point is the identity

$$\Delta(mp; u) = \Delta(m; u) + \Delta(m; u - \log p) \quad (p \nmid m). \tag{5.61}$$

When $L < k < K(n)$ we may apply this with $m = n_k$, $p = p_{k+1}(n)$, raising both sides to the power q, performing the binomial expansion on the right and the integration w.r.t. u over $(-\infty, \infty)$. We obtain

$$M_q(n_{k+1}) = 2M_q(n_k) + R_q(n_{k+1}) \tag{5.62}$$

where

$$R_q(n_{k+1}) := \sum_{j=1}^{q-1} \binom{q}{j} \int_{-\infty}^{\infty} \Delta(n_k; u)^{q-j} \Delta(n_k; u - \log p_{k+1}(n))^j \, du. \tag{5.63}$$

We want to average $R_q(n_{k+1})$ over 'normal' numbers n with fixed n_k and variable $p_{k+1}(n)$ in order to obtain an upper bound for $M_q(n_{k+1})$ in terms of n_k, p.p.x. To this end we first have to define a set A_k of integers a that are good 'candidates' to being the n_k of some normal n. We let A_k be the set of all integers a such that

(A_k1) $\mu(a)^2 = 1$, $\omega(a) = k - L$, $\log P^+(a) < e^{-L} \log x$,

(A_k2) $\log_2 a \leqslant \log_2 x - \frac{1}{2}L$,

(A_k3) $(1-\varepsilon)(j+L) < \log_2 p_j(a) < (1+\varepsilon)(j+L)$ $(1 \leqslant j \leqslant k-L)$,

where ε is a small positive number, fixed until the end of the proof. Set

$$A := \{n : n \leqslant x, n_k \in A_k \quad (L < k \leqslant K(n))\}.$$

Then card $A = (1 + o(1))x$. Indeed n_k satisfies (A_k1) trivially (in the specified range for k), (A_k2) follows p.p.x from Theorem 07, and (A_k3) is a consequence of the corollary to Theorem 10. Now put

$$S_k(x, a) := \text{card}\,\{n : n \leqslant x, n_k = a\}.$$

Lemma 56.2. *Let* $L < k \leqslant M$, $a \in A_k$, $P^+(a) < p < \exp\exp(\log_2 x - L)$. *Then*

$$S_{k+1}(x, ap) \ll \frac{x \log P^-(a)}{ap \log P^+(a)}.$$

Proof. Let b denote generically an integer such that $\omega(b) = L$, $\mu(b)^2 = 1$, $P^+(b) < P^-(a)$. Then $S_{k+1}(x, ap)$ counts the integers $n = bapdm \leqslant x$ where all prime factors of d divide bap, and $P^-(m) > p$. Hence

$$S_{k+1}(x, ap) = \sum_b \sum_{t \leqslant x/bap} \chi(t)$$

where χ is the completely multiplicative function such that $\chi(p') = 1$ if $p' | bap$ or $p' > p$, $\chi(p') = 0$ else. The conditions imposed on b, a, p imply $bap = x^{o(1)}$ provided $L \to \infty$ suitably slowly: hence Theorem 01 implies that the t-sum above is $\ll x\varphi(ap)^{-1}\varphi(b)^{-1}(\log p)^{-1}$. We notice that (A_k3) implies that $\varphi(ap) \gg ap$ and use the trivial bound

$$\sum_b \varphi(b)^{-1} \leqslant \prod_{p' < P^-(a)} \left(1 + \frac{1}{p'-1}\right) \ll \log P^-(a)$$

which yields the required estimate.

We may now carry out the averaging process. We summarize the estimate which results as follows.

Lemma 56.3. *Define*

$$W_q(m) := \sum_{j=1}^{q-1} \binom{q}{j} M_j(m) M_{q-j}(m).$$

Then, for $k \in (L, M)$, $q \geqslant 2$, $a \in A_k$, *we have*

$$\sum_{\substack{n \in A \\ n_k = a \\ K(n) > k}} R_q(n_{k+1}) \ll \frac{x \log P^-(a)}{a \log^2 P^+(a)} W_q(a)\left(1 + 2^q \exp(-\sqrt{\log P^+(a)})\right). \tag{5.64}$$

Proof. We have

$$\begin{aligned} T_j(a) := & \sum_{\substack{n \in A \\ n_k = a \\ K(n) > k}} \int_{-\infty}^{\infty} \Delta(a; u)^{q-j} \Delta(a; u - \log p_{k+1}(n))^j \, du \\ & \leqslant \sum_{\substack{p > P^+(a) \\ ap \in A_{k+1}}} \int_{-\infty}^{\infty} \Delta(a; u)^{q-j} \Delta(a; u - \log p)^j \, du \cdot S_{k+1}(x, ap) \\ & \ll \frac{x \log P^-(a)}{a \log^2 P^+(a)} \int_{-\infty}^{\infty} \Delta(a; u)^{q-j} \sum_{p > P^+(a)} \Delta(a; u - \log p)^j \frac{\log p}{p} \, du \end{aligned}$$

by Lemma 56.2. We expand $\Delta(a; u - \log p)^j$ as a multiple sum, and invert summations. The sum over p involved is

$$\sum_{d_1, d_2, \ldots, d_j | a}^{*} \sum \left\{ \frac{\log p}{p} : \frac{e^u}{\min_i d_i} < p \leqslant \frac{e^{u+1}}{\max_i d_i}, p > P^+(a) \right\} \tag{5.65}$$

where the star denotes that summation is restricted to j-tuples of divisors $\{d_1, d_2, \ldots, d_j\}$ of a such that $\log(\max d_i / \min d_i) \leqslant 1$. Put $\alpha = \max(u - \log \min d_i, \log P^+(a))$, $\beta = u + 1 - \log \max d_i$. By the Prime Number Theorem, the sum over p above is $\beta - \alpha + O(\exp(-\sqrt{\alpha}))$, and we rearrange the main terms and add the remainders in (5.65) to obtain

$$\int_{\log P^+(a)}^{\infty} \Delta(a; u - v)^j \, dv + O(M_j^*(a) \exp(-\sqrt{\log P^+(a)})) \tag{5.66}$$

where

$$M_j^*(a) := \sum_{d_1, d_2, \ldots, d_j | a}^{*} 1 \leqslant 2^j M_j(a) \quad (a, j \in \mathbb{Z}^+). \tag{5.67}$$

To establish the inequality on the right, notice that

$$\begin{aligned} M_j^*(a) & \leqslant \sum_{d_1, d_2, \ldots, d_j | a} \left(2 - \log \frac{\max d_i}{\min d_i}\right)^+ = \int_{-\infty}^{\infty} (\Delta(a; u) + \Delta(a; u + 1))^j \, du \\ & \leqslant 2^{j-1} \int_{-\infty}^{\infty} \{\Delta(a; u)^j + \Delta(a; u + 1)^j\} \, du = 2^j M_j(a). \end{aligned}$$

We insert (5.67) into (5.66), observing that the main term in (5.66) does not exceed $M_j(a)$. We obtain

$$T_j(a) \ll \frac{x \log P^-(a)}{a \log^2 P^+(a)} M_{q-j}(a) M_j(a)(1 + 2^j \exp(-\sqrt{\log P^+(a)})).$$

Referring to (5.63) we see that this yields (5.64).

We may embark on the final stage of the proof. Since $M_q(n_{k+1}) \equiv M_q(n_k)$ when $k \geqslant K(n)$, Lemma 56.3 gives, for $k \in (L, M]$, $a \in A_k$, $q \geqslant 2$,

$$\sum_{\substack{n \in A \\ n_k = a}} (M_q(n_{k+1}) - 2M_q(n_k))^+ W_q(n_k)^{-1}$$
$$\ll xa^{-1} \exp\{(1+\varepsilon)L - 2(1-\varepsilon)k\}(1 + 2^q \exp(-(1.5)^k))$$

for $\varepsilon < \varepsilon_0$, where inequalities for $P^+(a)$, $P^-(a)$ have come from (A_k3). We sum this over $a \in A_k$, taking into account that

$$\sum_{a \in A_k} \frac{1}{a} \leqslant \prod_{\substack{(1-\varepsilon)L < \log_2 p \\ < (1+\varepsilon)k}} \left(1 + \frac{1}{p}\right) \ll \exp\{(1+\varepsilon)k - (1-\varepsilon)L\}$$

and we get

$$\sum_{n \in A} (M_q(n_{k+1}) - 2M_q(n_k))^+ W_q(n_k)^{-1} \ll x\mathrm{e}^{-(1-5\varepsilon)k}. \tag{5.68}$$

For each $k \in (L, M]$, the following set of inequalities hold (simultaneously), for all except $\ll x\mathrm{e}^{-\varepsilon k}$ of the integers $n \in A$:

$$M_q(n_{k+1}) \leqslant 2M_q(n_k) + \mathrm{e}^{-(1-7\varepsilon)k} W_q(n_k) \quad (1 \leqslant q \leqslant k). \tag{5.69}$$

After consideration of all the exceptions for different $k \in (L, M]$, we see that (5.69) holds p.p.x, simultaneously for all these k.

Let $1 < \delta < (1 - 7\varepsilon)/\log 2$. We are going to prove by induction on k that for each fixed $k \in (L, M]$

$$M_q(n_k) \leqslant 2^{\delta k} q! \quad (1 \leqslant q \leqslant k) \quad \text{p.p.x.} \tag{5.70}$$

For convenience put $e_1 = \exp(1 - 7\varepsilon)$. For $k = L + 1$ we have $M_q(n_k) \equiv 2$ for $q \geqslant 1$, $n \in A$, and (5.70) holds. Suppose it holds for k. When $q \leqslant k$, we may appeal to (5.69) to estimate $M_q(n_{k+1})$. This yields

$$M_q(n_{k+1}) \leqslant 2^{\delta(k+1)} q!(2^{1-\delta} + k(2^\delta/e_1)^k) \leqslant 2^{\delta(k+1)} q!$$

if L, and therefore k, is large enough. When $q = k + 1$ we can still use (5.70) to majorize $W_q(n_k)$ (which is a function of the $M_j(n_k)$, $j \leqslant q - 1 = k$), but we need some extra information to estimate $M_{k+1}(n_k)$ which is provided by Lemma 56.1 applied with $q = k$. Thus

$$M_{k+1}(n_k) \leqslant \Delta(n_k) M_k(n_k) \leqslant M_k(n_k)\left(1 + \frac{3}{e_1} M_k(n_k)^{1/k}\right),$$

because $(3/\mathrm{e})^{1+\varepsilon} < 3/e_1$. Whence, by (5.70) (with $q = k$),

$$M_{k+1}(n_k) \leqslant 2^{\delta(k+1)}(k+1)! \left\{\frac{2^{-\delta}}{k+1} + \frac{3}{e_1} \frac{(k!)^{1/k}}{k+1}\right\}$$
$$\leqslant 2^{\delta(k+1)}(k+1)! \left\{\frac{3}{\mathrm{e} \cdot e_1} + \mathrm{o}(1)\right\}$$

by Stirling's formula, and (5.69) gives

$$M_{k+1}(n_{k+1}) \leqslant 2^{\delta(k+1)}(k+1)!\left\{\frac{6}{\mathrm{e}\cdot e_1} + \mathrm{o}(1) + k(2^{\delta}/e_1)^k\right\}$$
$$\leqslant 2^{\delta(k+1)}(k+1)!$$

provided ε is suitably chosen, and k is large enough. This completes the induction and so proves (5.70). In particular we have

$$M_K(n_K) \leqslant 2^{\delta K} K! \quad \text{p.p.}x,$$

and (5.60) follows, on taking Kth roots and applying (5.56). This completes the proof of Theorem 56.

Notes on Chapter 5

§5.1. The concept of concentration function was introduced in 1937 by Paul Lévy as a tool for the study of sums of random variables. For any random variable X with distribution function F the concentration function is defined by the formula

$$Q(l) := \sup_{x \in \mathbb{R}} (F(x+l) - F(x)) \quad (l > 0).$$

An account of the subject may be found in the book of Hengartner and Theodorescu (1973). If $Q_n(l)$ denotes the concentration function of the r.v. D_n taking the values $\log d_i(n)$ with uniform probability $1/\tau(n)$, then

$$\Delta(n) = \tau(n) Q_n(1).$$

However the results available from Probability Theory do not go very far in the study of the Δ-function. The Kolmogorov–Rogozin inequality reads here

$$\Delta(n) \ll \tau(n)\omega(n)^{-1/2}, \tag{1}$$

as an immediate consequence of the canonical expansion as a sum of independent r.v.'s,

$$D_n = \sum_{p^\nu \| n} D_{p^\nu}.$$

The example following (4.65) shows that (1) is optimal (apart from the constant) and it seems that no finer general result has been obtained so far that could, for instance, yield a better p.p. estimate than (1), using information about the prime factor distribution of normal numbers.

The first proof of Erdős' conjecture appears in Maier & Tenenbaum (1984), where it is shown that the upper bound (5.13) for $E(n)$ holds p.p.x when $\psi \to \infty$ – the cardinality of the exceptional set is not estimated. As we mentioned in the remarks following Theorem 53, $R_1(x, \psi(x)) = o(1)$ only if $\psi(x) > c\sqrt{\log_3 x}$: this limits (5.14) as a p.p. lower bound for $E(n)$ and a slightly better result can be found in Erdős & Hall (1979): for every $\varepsilon > 0$ we have

$$E(n) > (\log n)^{1-\log 3} \exp\{-(1+\varepsilon)\log 3 \cdot \sqrt{(2\log_2 n \cdot \log_4 n)}\}. \tag{2}$$

If the Fourier transform method could somehow be applied to the problem of the p.p. upper bound for $\nabla(n, t)$, it is likely that the factor $\sqrt{\log_4 n}$ above could be replaced by any function tending to infinity.

The p.p.x lower bound (5.11) for $U(n, \alpha)$ was conjectured in Hall & Tenenbaum (1981).

§5.2. The main idea of the proof (that is the inequalities (5.24)) is essentially that of Erdős & Hall (1979).
§5.3. Theorem 56 is due to Maier & Tenenbaum (1985). The proof presented here is almost identical to the one in their paper.

It is easy to see that the upper asymptotic density of those integers which have two prime factors p, q such that $p < q < 2p$ is less than 1. Several questions about the propinquity of prime factors were discussed by Erdős (1959).

Exercises on Chapter 5

50. Let $\mathscr{E}$ denote the set of 'Erdős primitive' numbers, that is numbers having two divisors d, d' such that $d < d' \leqslant 2d$ but such that no proper divisor has this property.

(i) Use Theorem 21 to show that

$$\sum\{1: n \leqslant x, n \in \mathscr{E}\} \ll x(\log x)^{-\delta}(\log\log x)^{-1/2}$$

(with δ as in (2.3)).

(ii) Let $\mathscr{B} = \mathscr{B}(\mathscr{E})$ be the set of multiples of $\mathscr{E}$. Show that Behrend's Inequality (see Exercise 21) implies that $\underline{\mathbf{d}}\mathscr{B} \leqslant 1 - \prod\{(1 - 1/n): n \in \mathscr{E}\}$. Deduce from Theorem 52 that

$$\sum\{1/n: n \in \mathscr{E}\} = +\infty. \tag{1}$$

51. (Sequel to Exercise 50). Let $a_1 < a_2 < \cdots$ be the sequence of those integers having no divisor in $\mathscr{E}$. Show that, if (1) were false, then $\{a_j\}$ would have positive lower asymptotic density. Show that all numbers $n = pa_i$, with $(a_i, 6) = 1$, $p < a_i \leqslant 2p$, belong to $\mathscr{E}$, and deduce from this a contradiction – i.e. a direct proof of (1). [This argument, due to Erdős, was known to him long before the proof of his conjecture.]

52. Define $h(n) := \sum\{d_j(n)/d_{j+1}(n): 1 \leqslant j < \tau(n), (d_j(n), d_{j+1}(n)) = 1\}$. Show that $h(n) \geqslant c_0 \log_3 n$ p.p. What value of c_0 can you obtain? [Hint: see the proof of Theorem 55.]

53. Let $F(\alpha)$ denotes the distribution function of $\tau^\dagger(n)/\tau(n)$ (cf. Theorem 49), that is, for $\alpha \in \mathscr{C}F$, the asymptotic density of the sequence $\{n \in \mathbb{Z}^+: \tau^\dagger(n) \leqslant \alpha\tau(n)\}$. By applying Theorem 51 with $u = \frac{3}{2}$ and v large but fixed, prove that $1 \in \mathscr{C}F$.

54. Define $T^*(n) = T(n, 0) - \tau(n)$, and show that, for all n, $T^*(n) \geqslant \tau(n) - \tau^+(n)$. Let F^* denote the distribution function of $T^*(n)/\tau(n)$ (cf. §4.5) and deduce from Exercise 53 that $0 \in \mathscr{C}F^*$.

55. Let $\varepsilon \in (0, 1)$ be fixed and define

$$\operatorname{Prop}(n) := \operatorname{card}\{j: 1 \leqslant j < \tau(n),\ d_{j+1}(n) < (1 + \varepsilon)d_j(n)\}.$$

Show that $\operatorname{Prop}(n)/\tau(n)$ has a distribution function which is continuous at 0 and 1.

6

Hooley's Δ_r-functions – sharp bounds

6.1 Introduction

We have already met the function $\Delta(n)$ in Chapters 4 & 5: it is the maximum number of divisors of n contained in any interval of logarithmic length 1. More generally, we set

$$\Delta(n; u_1, u_2, \ldots, u_{r-1}) = \text{card}\,\{d_1 d_2 \cdots d_{r-1} | n : u_i < \log d_i \leqslant u_i + 1 \quad (1 \leqslant i < r)\} \tag{6.1}$$

so that the definition of the Δ_r-function in Hooley (1979) is equivalent to

$$\Delta_r(n) = \max_{u_1, u_2, \ldots, u_{r-1}} \Delta(n; u_1, u_2, \ldots, u_{r-1}), \tag{6.2}$$

in particular $\Delta_2(n) = \Delta(n)$. It is convenient to define $\Delta_1(n) \equiv 1$.

We note that $\Delta_r(n)$ is a non-decreasing function of r: we have only to restrict the range for the maximum in (6.2) by setting $u_{r-1} = 0$, and take $d_{r-1} = 1$. Also, for fixed r the renormalized function $2^{1-r}\Delta_r(n)$ is *supermultiplicative*, that is for relatively prime m and n we have $2^{1-r}\Delta_r(mn) \geqslant (2^{1-r}\Delta_r(m))(2^{1-r}\Delta_r(n))$. For if $d_1 d_2 \ldots d_{r-1} | m$, $t_1 t_2 \ldots t_{r-1} | n$, $u_i < \log d_i \leqslant u_i + 1$, $v_i < \log t_i \leqslant v_i + 1$ $(1 \leqslant i < r)$, then $(d_1 t_1)(d_2 t_2) \ldots (d_{r-1} t_{r-1}) | mn$, and moreover $u_i + v_i < \log d_i t_i \leqslant u_i + v_i + 2$ $(1 \leqslant i < r)$. Hence

$$\Delta(m; u_1, u_2, \ldots, u_{r-1})\Delta(n; v_1, v_2, \ldots, v_{r-1}) \leqslant \sum_{\varepsilon} \Delta(mn; u_1 + v_1 + \varepsilon_1, \ldots, u_{r-1} + v_{r-1} + \varepsilon_{r-1}) \tag{6.3}$$

where the sum is over the 2^{r-1} vectors $\varepsilon = (\varepsilon_1, \varepsilon_2, \ldots, \varepsilon_{r-1}) \in \{0, 1\}^{r-1}$. The result follows on taking the maximum on the left of (6.3). Often $\Delta_r(mn)$ is larger than $\Delta_r(m)\Delta_r(n)$. For example, we have $\Delta(2^\alpha) = 2$ $(\alpha \geqslant 1)$, $\Delta(p^\alpha) = 1$ $(p \geqslant 3, \alpha \geqslant 1)$. But $\Delta(2^\alpha 3^\beta) \geqslant \beta + 1$ whenever $2^\alpha > 3^\beta$. To see this, set $u_1 = \beta \log 3$. For each b, $0 \leqslant b \leqslant \beta$, there exists a, $0 \leqslant a \leqslant \alpha$, such that $d := 2^a 3^b$ satisfies $u_1 < \log d \leqslant u_1 + 1$. The reader may find it helpful to evaluate $\Delta_r(n)$ in a few simple cases (cf. Exercise 60).

The function $\Delta(n)$ is a measure of the propinquity of divisors: it is also a concentration function in the sense of Paul Lévy (1937). At first sight the extension to $\Delta_r(n)$ seems a little artificial; however, the function presents itself in a natural way in Hooley's 'New Technique' (1979). This is a method which is effective in a wide range of problems in the areas of Diophantine Approximation and Waring's Problem. What is needed in these applications is an upper bound for the sum

$$S_r(x) := \sum_{n \leqslant x} \Delta_r(n) \tag{6.4}$$

appreciably better than what would be obtained from the trivial inequality $\Delta_r(n) \leqslant \tau_r(n)$. Since $\Delta_r(n) \geqslant 1$, and the average order of $\tau_r(n)$ is $(\log n)^{r-1}/(r-1)!$, we can of course only save a power of a logarithm here. However there are circumstances where this is exactly what is needed. For example Hooley used his upper bound $S_2(x) \ll x(\log x)^{4/\pi - 1}$ to show that the number of integers $n \leqslant x$ expressible in the form $n = a^2 + b^4 + c^4$ is $\gg_\varepsilon x/(\log x)^{4/\pi - 1 + \varepsilon}$. The results of Chapter 7 enable us to replace the denominator here by an (explicit) slowly oscillating function of $\log x$ – see Tenenbaum (1986). Generally, Hooley's technique is at its best when the result envisaged is sharp except for a power of a logarithm, or when such a saving in an error term is the only way to obtain an asymptotic formula. Vaughan (1986, 1986a) is the most recent application at the time of writing.

We consider a more general sum than $S_r(x)$, namely

$$S_r(x, y) := \sum_{n \leqslant x} \Delta_r(n) y^{\omega(n)}. \tag{6.5}$$

Definition. *For $r \geqslant 2$, $y > 0$ we denote by $\alpha(r, y)$ the infimum of the numbers ξ such that $S_r(x, y) \ll_\xi x(\log x)^\xi$.*

By applying Hölder's inequality to (6.5) we see that $\alpha(r, y)$ is a convex function of $\log y$: as such it is continuous. In Chapter 7 (cf. Theorems 70, 71) we shall prove

$$\alpha(r, y) = y - 1 + (r-1)(y-1)^+ \quad (r \geqslant 2, y > 0). \tag{6.6}$$

The value $y = 1$ emerges as quite special independently of its importance in the New Technique and $S_r(x)$ itself is better understood within this slightly more general setting.

6.2 Lower bounds

We begin with a simple lower bound for $S_r(x, y)$ which helps to fix our ideas; it also serves as a target for our upper bounds. Let $k = [\log n] + 1$. Then

$$\begin{aligned}\tau_r(n) &= \sum_{0 \leqslant h_1 < k} \cdots \sum_{0 \leqslant h_{r-1} < k} \Delta(n; h_1, h_2, \ldots, h_{r-1}) \\ &\leqslant k^{r-1} \Delta_r(n) \leqslant (\log en)^{r-1} \Delta_r(n).\end{aligned}$$

Theorem 60. *For $r \geqslant 2$, $y > 0$ we have*

$$\Delta_r(n) \geqslant \max(1, \tau_r(n)/(\log en)^{r-1}), \tag{6.7}$$

$$S_r(x, y) \gg_{r,y} x(\log x)^{y-1+(r-1)(y-1)^+}(\log_2 x)^{\delta(y,1)}, \tag{6.8}$$

$$\alpha(r, y) \geqslant y - 1 + (r-1)(y-1)^+. \tag{6.9}$$

Proof. We have to prove (6.8). For $r \geqslant 2$, $y > 0$ there exists a constant $C_r(y) > 0$ such that

$$\sum_{n \leqslant x} \tau_r(n) y^{\omega(n)} \sim C_r(y) x (\log x)^{ry-1},$$

this may be proved by the usual analytic technique starting from Perron's formula. Together with (6.7) it yields (6.8) provided $y \neq 1$.

Erdős had an unpublished proof that $S_2(x)/x \to \infty$ and this was reproduced by Hooley (1979). The method used can be made to give a specific, but too slow, rate of growth. Instead we employ the following result – we recall the definition (4.1) of the function $T(n, \alpha)$.

Lemma 60.1

$$\Delta(n) \geqslant T(n, 0)/2\tau(n) \quad (n \geqslant 1). \tag{6.10}$$

Proof. By definition

$$\begin{aligned} T(n, 0) &= \operatorname{card}\{d_1, d_2 : d_1 | n, d_2 | n, |\log(d_1/d_2)| \leqslant 1\} \\ &= \sum_{d_1 | n} (\Delta(n; \log d_1 - 1) + \Delta(n; \log d_1)) \\ &\leqslant 2\tau(n)\Delta(n). \end{aligned}$$

A corollary is that $\Delta(n) \geqslant \frac{1}{2} T(n, 0)/2^{\Omega(n)}$, and (6.8) (with $y = 1$) follows from (4.9). For $r > 2$ we just use the inequality $\Delta_r(n) \geqslant \Delta(n)$.

6.3 The critical interval

The title of the present chapter arises from the fact that for certain values of y the lower bound (6.8) is actually sharp, apart from constants; that is we can replace $\gg_{r,y}$ in (6.8) by $\asymp_{r,y}$. More precisely, we are going to show that there exist non-empty intervals $0 < y < y'$, $y'' < y < \infty$ on which

$$S_r(x, y) \asymp_{r,y} x(\log x)^{y-1+(r-1)(y-1)^+}. \tag{6.11}$$

Definition. Λ_r^- *is the supremum of the numbers y' such that* (6.11) *holds for each $y \in (0, y')$. Λ_r^+ is the infimum of the numbers y'' such that* (6.11) *holds for each $y \in (y'', \infty)$.*

Notice that these bounds are attained. Also, from Theorem 60, $\Lambda_r^- \leqslant 1 \leqslant \Lambda_r^+$.

We show in Theorem 62 that if $\Lambda_r^- < \Lambda_r^+$ *then* (6.11) *is false throughout* $(\Lambda_r^-, \Lambda_r^+)$.

Definition. *For each* $r \geqslant 2$, *the critical interval is the non-empty interval* $[\Lambda_r^-, \Lambda_r^+]$.

The scheme of this and the next chapter is as follows. In the remainder of this chapter we prove the existence of Λ_r^-, Λ_r^+ and establish bounds for these functions. In Chapter 7 we give upper bounds for $S_r(x, y)$ in the critical interval which are sharp except for factors involving certain explicit slowly oscillating functions of $\log x$. A corollary is (6.6), the formula for $\alpha(r, y)$.

6.4 Technical preparation

At several points of our treatment of upper bounds for $S_r(x, y)$ in this and the next chapter, it will be convenient to restrict our attention to squarefree numbers, moreover to deal with the weighted sum

$$Z_r(x, y) := \sum_{n \leqslant x} \frac{\mu^2(n)}{n} \Delta_r(n) y^{\omega(n)}. \tag{6.12}$$

Theorem 61. *For each* $r \geqslant 2$ *and* $y > 0$ *we have*

$$S_r(x, y) \ll_{r,y} \frac{x}{\log x} Z_r(x, y). \tag{6.13}$$

Lemma 61.1. *For all integers* m, n *(not necessarily coprime) we have*

$$\Delta_r(mn) \leqslant \tau_r(m) \Delta_r(n). \tag{6.14}$$

Proof of lemma. Every divisor of mn can be written in the form dt, where $d|m$, $t|n$. Hence

$$\begin{aligned}\Delta(mn; u_1, \ldots, u_{r-1}) &\leqslant \sum_{d_1 \ldots d_{r-1} | m} \Delta(n; u_1 - \log d_1, \ldots, u_{r-1} - \log d_{r-1}) \\ &\leqslant \tau_r(m) \Delta_r(n).\end{aligned}$$

The result follows on maximizing the left-hand side.

Proof of Theorem 61. Every integer n may be written in the form d^2k where k is squarefree. Put $y_1 = \max(y, 1)$. Then

$$\begin{aligned}\sum_{n \leqslant x} \Delta_r(n) \frac{y^{\omega(n)}}{n} &\leqslant \sum_{d=1}^{\infty} \tau_r(d^2) \frac{y_1^{\omega(d)}}{d^2} \sum_{k \leqslant x} \frac{\mu^2(k)}{k} \Delta_r(k) y^{\omega(k)} \\ &\ll_{r,y} Z_r(x, y)\end{aligned}$$

because the infinite series on the right is convergent. Next,

$$S_r(x,y)\log x=\sum_{n\leqslant x}\Delta_r(n)y^{\omega(n)}\log\frac{x}{n}+\sum_{n\leqslant x}\Delta_r(n)y^{\omega(n)}\log n,$$

and since $\log(x/n)<x/n$ the first sum on the right is $\ll_{r,y}xZ_r(x,y)$. Also

$$\sum_{n\leqslant x}\Delta_r(n)y^{\omega(n)}\log n=\sum_{n\leqslant x}\Delta_r(n)y^{\omega(n)}\sum_{d\mid n}\Lambda(d)$$

and we write $n=md$, invert the order of summation on the right, and employ Lemma 61.1 again. The sum above is

$$\begin{aligned}&\leqslant\sum_{d\leqslant x}\Lambda(d)\sum_{m\leqslant x/d}\tau_r(d)\Delta_r(m)y_1y^{\omega(m)}\\&\leqslant y_1\sum_{m\leqslant x}\Delta_r(m)y^{\omega(m)}\sum_{d\leqslant x/m}\Lambda(d)\tau_r(d)\end{aligned}$$

and the inner sum on the right is $\ll_r x/m$. So we have $\ll_{r,y}xZ_r(x,y)$ on the right. Putting these inequalities together, we get $S_r(x,y)\log x\ll_{r,y}xZ_r(x,y)$ which is (6.13).

Lemma 62.1. *For every $r\geqslant 2$ and $0<y<z$ we have*

$$Z_r(x,z)\ll_{r,y,z}Z_r(x,y)(\log x)^{rz-ry},$$

and, for $0<z<y\leqslant 4$,

$$Z_r(\sqrt{x},z)\ll_{r,y,z}(\log x)^{z-y}\sum_{n\leqslant x}\Delta_r(n)\frac{y^{\omega(n)}}{n}.$$

Proof. We begin with the first part. We have

$$Z_r(x,z)=\sum_{n\leqslant x}\frac{\mu^2(n)}{n}\Delta_r(n)y^{\omega(n)}\sum_{d\mid n}\mu^2(d)\left(\frac{z}{y}-1\right)^{\omega(d)}$$

and we write $n=md$, $\Delta_r(n)\leqslant\tau_r(d)\Delta_r(m)$. Hence

$$Z_r(x,z)\leqslant\sum_{d\leqslant x}\frac{\mu^2(d)}{d}\tau_r(d)(z-y)^{\omega(d)}\sum_{m\leqslant x}\frac{\mu^2(m)}{m}\Delta_r(m)y^{\omega(m)}$$

and the first sum on the right is $\ll_{r,y,z}(\log x)^{rz-ry}$. This is the result stated. Next, let $z<y\leqslant 4$, so that $z(y-z)\leqslant y$. Then

$$\begin{aligned}&Z_r(\sqrt{x},z)(\log x)^{y-z}\\&\ll_{r,y,z}\sum_{m\leqslant\sqrt{x}}\frac{\mu^2(m)}{m}\Delta_r(m)z^{\omega(m)}\sum_{d\leqslant\sqrt{x}}\frac{\mu^2(d)}{d}(y-z)^{\omega(d)}\\&\ll_{r,y,z}\sum_{n\leqslant x}\frac{1}{n}\Delta_r(n)\sum_{m\mid n}\mu^2(m)\mu^2\left(\frac{n}{m}\right)z^{\omega(m)}(y-z)^{\omega(n/m)},\end{aligned}$$

on putting $md=n$ and $\Delta_r(m)\leqslant\Delta_r(n)$. The inner sum is multiplicative and for $n = p^\alpha$ it equals y if $\alpha = 1$, $z(y-z) \leqslant y$ if $\alpha = 2$, and 0 if $\alpha \geqslant 3$. So it is $\leqslant y^{\omega(n)}$ in general, and this is the required result.

Theorem 62. *If* $\Lambda_r^- < \Lambda_r^+$ *then* (6.11) *is false throughout* $(\Lambda_r^-, \Lambda_r^+)$.

Proof. Let $y > 1$ and (6.11) hold. By partial summation, $Z_r(x, y) \ll (\log x)^{ry-r+1}$, and so, by the first part of Lemma 62.1 and (6.13), (6.11) holds for $z > y$. Hence $y \geqslant \Lambda_r^+$. If $y < 1$ and (6.11) holds, it holds for $z < y$ by a similar argument, using the second part of Lemma 62.1. Therefore $y \leqslant \Lambda_r^-$.

6.5 Iteration inequalities

An elementary approach to upper bounds for $S_r(x, y)$ is based on the following inequalities.

Lemma 63.1

$$\Delta(n; u)^2 \leqslant \sum_{\delta\delta' | n}{}^{*} \Delta(n/\delta\delta'; u - \log \delta), \tag{6.15}$$

*where * denotes that summation is restricted by the conditions* $|\log(\delta/\delta')| < 1$, $(\delta, \delta') = 1$.

Proof. Let $\Delta(n; u) = k$, and $d_1, d_2, \ldots, d_k \in (\mathrm{e}^u, \mathrm{e}^{u+1}]$ be the divisors of n in this range. For each of the k^2 choices of $i, j \leqslant k$ write $\delta_{ij} = d_i/(d_i, d_j)$, $\delta'_{ij} = d_j/(d_i, d_j)$, so that $|\log(\delta_{ij}/\delta'_{ij})| < 1$, $(\delta_{ij}, \delta'_{ij}) = 1$. If we now sum over all possible δ, δ' allowed by the summation conditions in (6.15) we shall account for all k^2 choices of i, j if we reckon how many pairs can give rise to the same δ, δ', that is have $\delta_{ij} = \delta$, $\delta'_{ij} = \delta'$. Let $d_i = \delta t$, $d_j = \delta' t$. Then t is restricted in two ways. Firstly $t | n/\delta\delta'$, secondly $u - \log \delta < t \leqslant u - \log \delta + 1$. (We ignore the similar condition involving δ'.) The number of these t is $\Delta(n/\delta\delta'; u - \log \delta)$, and this proves (6.15). A corollary is

$$\Delta(n)^2 \leqslant \sum_{\delta\delta' | n}{}^{*} \Delta(n/\delta\delta'). \tag{6.16}$$

This is the case $r = 2$ of the following result.

Theorem 63. *For* $r \geqslant 2$ *and all* n, *we have*

$$\Delta_r(n)^2 \leqslant \Delta_{r-1}(n) \sum_{\delta\delta' | n}{}^{*} \Delta_r(n/\delta\delta'). \tag{6.17}$$

*where * has the meaning defined in the previous lemma.*

Proof. Choose $v_1, v_2, \ldots, v_{r-1}$ so that $\Delta(n; v_1, \ldots, v_{r-1}) = \Delta_r(n)$. Then we have (we may assume $r \geqslant 3$)

$$\Delta_r(n) = \sum_{\substack{d_1 \ldots d_{r-2} | n \\ v_i < \log d_i \leqslant v_i + 1 \quad (i \leqslant r-2)}} \Delta(n/d_1 d_2 \ldots d_{r-2}; v_{r-1}),$$

whence by the Cauchy–Schwarz inequality

$$\Delta_r(n)^2 \leqslant \Delta_{r-1}(n) \sum_{\substack{d_1 \ldots d_{r-2} | n \\ v_i < \log d_i \leqslant v_i + 1 \quad (i \leqslant r-2)}} \Delta(n/d_1 d_2 \ldots d_{r-2}; v_{r-1})^2.$$

We apply (6.15) to the summand, and invert the resulting double sum. This yields

$$\Delta_r(n)^2 \leqslant \Delta_{r-1}(n) \sum_{\delta\delta' | n}{}^{*} \sum_{\substack{d_1 d_2 \ldots d_{r-2} | n/\delta\delta' \\ v_i < \log d_i \leqslant v_i + 1 \quad (i \leqslant r-2)}} \Delta(n/\delta\delta' d_1 \ldots d_{r-2}; v_{r-1} - \log \delta).$$

But the inner sum is $\Delta(n/\delta\delta'; v_1, \ldots, v_{r-2}, v_{r-1} - \log \delta)$ which does not exceed $\Delta_r(n/\delta\delta')$. This gives (6.17).

6.6 Small y – the lower bound for Λ_r^-

Theorem 64. *For every $r \geqslant 2$ we have $\Lambda_r^- \geqslant \frac{1}{2}$. More precisely, for $0 < y \leqslant \frac{1}{2}$ we have*

$$S_r(x, y) \ll_{r,y} x(\log x)^{y-1} (\log_2 x)^{(r-1)\delta(2y,1)}. \tag{6.18}$$

Proof. We work with $Z_r(x, y)$, relating this to $Z_{r-1}(x, y)$ via the inequality (6.17) to set up an induction over r. Let $r \geqslant 2$. By Cauchy–Schwarz,

$$Z_r(x, y)^2 \leqslant \sum_{n \leqslant x} \frac{\mu^2(n)}{n} \Delta_{r-1}(n) y^{\omega(n)} \sum_{n \leqslant x} \frac{\mu^2(n)}{n} \frac{\Delta_r(n)^2}{\Delta_{r-1}(n)} y^{\omega(n)},$$

and we notice that the first sum on the right is $Z_{r-1}(x, y)$. We apply (6.17) to the second sum, which does not exceed

$$\sum_{n \leqslant x} \frac{\mu^2(n)}{n} y^{\omega(n)} \sum_{\delta\delta' | n}{}^{*} \Delta_r(n/\delta\delta') \leqslant \sum_{\substack{\delta\delta' \leqslant x \\ |\log(\delta/\delta')| < 1}} \frac{\mu^2(\delta\delta')}{\delta\delta'} y^{\omega(\delta\delta')} Z_r\left(\frac{x}{\delta\delta'}, y\right).$$

We make the trivial estimate $Z_r(x/\delta\delta', y) \leqslant Z_r(x, y)$. Of course such a step applied to S_r would be quite useless. We divide through by $Z_r(x, y)$ so that we have

$$\begin{aligned} Z_r(x, y) &\leqslant Z_{r-1}(x, y) \sum_{\substack{\delta\delta' \leqslant x \\ |\log(\delta/\delta')| < 1}} \frac{\mu^2(\delta\delta')}{\delta\delta'} y^{\omega(\delta\delta')} \\ &\leqslant Z_{r-1}(x, y) \sum_{\delta < e\sqrt{x}} \frac{y^{\omega(\delta)}}{\delta} \sum_{\delta/e < \delta' < \delta e} \frac{y^{\omega(\delta')}}{\delta'} \\ &\ll_y Z_{r-1}(x, y) \sum_{\delta < e\sqrt{x}} \frac{y^{\omega(\delta)}}{\delta} (\log \delta e)^{y-1}. \end{aligned}$$

We estimate the sum on the right by partial summation: it is $\ll_y (\log x)^{(2y-1)^+} (\log_2 x)^{\delta(2y,1)}$. For $y \leqslant \frac{1}{2}$ this gives

$$Z_r(x, y) \ll_y Z_{r-1}(x, y)(\log_2 x)^{\delta(2y,1)}$$

and, since $Z_1(x, y) \ll_y (\log x)^y$ for all y, we obtain, by induction on r,

$$Z_r(x, y) \ll_{r,y} (\log x)^y (\log_2 x)^{(r-1)\delta(2y,1)} \quad (r \geqslant 2, y \leqslant \tfrac{1}{2}).$$

We apply Theorem 61 to complete the proof.

This method works with partial success and can be refined in various ways when $y > \frac{1}{2}$. First, in the application of the Cauchy–Schwarz inequality we can replace the factor $y^{\omega(n)}$ which appears in both sums on the right by $(y^2/z)^{\omega(n)}$ and $z^{\omega(n)}$ respectively, where z may then be chosen optimally. Second, there is a more general inequality than (6.17) in which the factor $\Delta_{r-s+1}(n)$ appears outside the sum, with $s \geqslant 2$ at our disposal. The sum on the right of (6.17) becomes somewhat more complicated. These ideas were exploited in Hall & Tenenbaum (1982, 1984). They lead to the *iteration inequality*

$$2\alpha(r, y) \leqslant \alpha(r, z) + (s-1)(sz-1)^+ + \alpha(r-s+1, y^2/z), \tag{6.19}$$

valid for $1 \leqslant s \leqslant r$, y, $z > 0$ – there is a more precise version involving $Z_r(x, y)$, $Z_r(x, z)$, $Z_{r-s+1}(x, y^2/z)$ and a power (depending on the various parameters) of $\log_2 x$; and this yields bounds for $\alpha(r, y)$ or $Z_r(x, y)$ in the critical interval by means of an iterative procedure in which s and z have to be chosen optimally at each stage. In its present form this method has been superseded by that of Chapter 7, which evaluates $\alpha(r, y)$. However, it is possible that the argument leading to (6.19) can be refined to yield a deeper inequality, so that the iteration method would not only evaluate $\alpha(r, y)$ but give new, finer, estimates for $S_r(x, y)$ in the critical interval.

6.7 Fourier transforms – initial treatment

We now turn to a method, rooted in the ideas of Chapter 3, which is very successful in the study of the sum $S_r(x, y)$ when y is rather larger than 1.

We begin with a result due to Hooley (1979), generalized by the simple insertion of the parameter y.

Theorem 65 (C. Hooley). *Let $r \geqslant 2$, $y > 0$. Then*

$$S_r(x,y) \ll_{r,y} \frac{x}{L}\int_{-1}^{1}\cdots\int_{-1}^{1} \cdot\exp\left\{y\int_{1}^{L}|1+e^{i\theta_1 t}+e^{i\theta_2 t}+\cdots+e^{i\theta_{r-1}t}|\frac{dt}{t}\right\}d\theta_1\cdots d\theta_{r-1} \tag{6.20}$$

where $L = \log x$.

Proof. We bound $\Delta_r(n)$ from above by a multiple integral involving a multiplicative function of n. Put

$$w(u) = \left(\frac{\sin\frac{1}{2}u}{\frac{1}{2}u}\right)^2 = \int_{-1}^{1}(1-|\theta|)e^{i\theta u}\,d\theta. \tag{6.21}$$

Then $w(u) \geqslant 0$ for real u, $w(u) \geqslant w(1) > 0$ for $|u| \leqslant 1$. Hence

$$\Delta(n; u_1, u_2, \ldots, u_{r-1}) \ll_r \sum_{d_1 d_2 \ldots d_{r-1}|n}\ \prod_{j<r} w(\log d_j - u_j).$$

We insert the integral representation (6.21) for each factor in the product on the right, and reverse the order of summation and integration. The integral involves the function

$$\tau(n; \theta_1, \theta_2, \ldots, \theta_{r-1}) := \sum_{d_1 d_2 \ldots d_{r-1}|n} d_1^{i\theta_1} d_2^{i\theta_2} \ldots d_{r-1}^{i\theta_{r-1}}, \tag{6.22}$$

which is of course a generalization of $\tau(n,\theta)$ as in (3.1). We deduce that

$$\Delta_r(n) \ll_r \int_{-1}^{1}\cdots\int_{-1}^{1}|\tau(n;\theta_1,\theta_2,\ldots,\theta_{r-1})|\,d\theta_1\ldots d\theta_{r-1}. \tag{6.23}$$

The insertion of modulus signs removes the dependence on $u_1, u_2, \ldots, u_{r-1}$ in the integrand and so provides an upper bound for the maximum value of the integral – this step occurs in the theory of concentration functions in a similar fashion. Unfortunately it is the Achilles heel of the method as a means of estimating $S_r(x, 1)$. This point will become clearer as we proceed.

When n is squarefree, we have

$$\tau(n; \theta_1, \ldots, \theta_{r-1}) = \prod_{p|n}(1 + p^{i\theta_1} + p^{i\theta_2} + \cdots + p^{i\theta_{r-1}})$$

and we insert the weight $y^{\omega(n)}/n$ and sum, to obtain

$$Z_r(x,y) \ll_r \int_{-1}^{1}\cdots\int_{-1}^{1}\prod_{p\leqslant x}\left(1+\frac{y}{p}|1+p^{i\theta_1}+\cdots+p^{i\theta_{r-1}}|\right)d\theta_1\ldots d\theta_{r-1}$$

and then, via (6.13) and the inequality $1+u\leqslant e^u$,

$$S_r(x,y)\ll_r \frac{x}{L}\int_{-1}^{1}\cdots\int_{-1}^{1}\exp\left\{y\sum_{p\leqslant x}\frac{|1+p^{i\theta_1}+\cdots+p^{i\theta_{r-1}}|}{p}\right\}d\theta_1\ldots d\theta_{r-1}.$$

It remains to relate the sum over p above to the integral w.r.t. t in (6.20). We write

$$f(\tau):=\frac{1}{\tau}|1+\tau^{i\theta_1}+\tau^{i\theta_2}+\cdots+\tau^{i\theta_{r-1}}|\quad(\tau>0),$$

so that f is differentiable except at the isolated points where it vanishes. We have

$$\sum_{p\leqslant x}f(p)=\pi(x)f(x)-\int_2^x\pi(\tau)f'(\tau)\,d\tau,$$

and we employ the Prime Number Theorem in the form $\pi(\tau)-\operatorname{li}\tau\ll\tau/\log^2\tau\ (\tau>1)$. Together with the inequalities $f(\tau)\leqslant r/\tau$, $|f'(\tau)|\leqslant(r+|\theta_1|+\cdots+|\theta_{r-1}|)/\tau^2<2r/\tau^2$, this yields

$$\sum_{p\leqslant x}f(p)=\int_1^L e^t f(e^t)\frac{dt}{t}+O(r),$$

and (6.20) follows.

It is important to understand why (6.20) is weak when y is too small. Let I denote the multiple integral in (6.20) with the ranges restricted to $0\leqslant\theta_i\leqslant 1\ (1\leqslant i<r)$. The logarithmic function is concave and Jensen's inequality gives

$$\log I\geqslant\int_0^1\cdots\int_0^1 y\int_1^L|1+e^{i\theta_1 t}+\cdots+e^{i\theta_{r-1}t}|\frac{dt}{t}d\theta_1\ldots d\theta_{r-1}.$$

We integrate w.r.t. t last. For t fixed, substitute $\theta_j=\varphi_j/t\ (1\leqslant j<r)$ and set $N=[t/2\pi]$. Then

$$\begin{aligned}\log I&\geqslant y\int_1^L t^{-r}dt\int_0^{2\pi N}\cdots\int_0^{2\pi N}|1+e^{i\varphi_1}+\cdots+e^{i\varphi_{r-1}}|d\varphi_1\ldots d\varphi_{r-1}\\&\geqslant y\int_1^L t^{-r}(2\pi N)^{r-1}H_r\,dt\geqslant yH_r(\log L+O(r)),\end{aligned}$$

where

$$H_r=\frac{1}{(2\pi)^{r-1}}\int_0^{2\pi}\cdots\int_0^{2\pi}|1+e^{i\varphi_1}+\cdots+e^{i\varphi_{r-1}}|d\varphi_1\cdots d\varphi_{r-1}.\quad(6.24)$$

Hence $I\gg_{r,y}L^{H_r y}$ and the right-hand side of (6.20) is

$$\gg_{r,y}x(\log x)^{H_r y-1}.$$

The integral (6.24) is familiar in the theory of random walks. We may deduce from the Central Limit Theorem that $H_r \sim \frac{1}{2}\sqrt{(\pi r)}$; what is more relevant now is that for every $r \geqslant 2$ we have $H_r > 1$ (cf. Exercise 62). It follows that when $y \leqslant 1$ an upper bound for $S_r(x, y)$ derived from (6.20), far from being sharp, will have the wrong exponent of $\log x$. Indeed $\alpha(r, y) = y - 1 + (r-1)(y-1)^+ < H_r y - 1$ for $y < (r-1)/(r-H_r)$. For these values of y the method is inefficient and we can trace the weakness back to the introduction of the modulus signs in (6.23). A rather surprising method of improving (6.23) (in certain applications) used this weakness (Hall & Tenenbaum (1984)): for $\lambda \geqslant 0$ it is easy to see that

$$\Delta(n) \ll \int_{-1}^{1} \prod_{p \mid n} |\lambda p^{-i\theta} + 1 + p^{i\theta} + \lambda p^{2i\theta}| \, d\theta$$

for squarefree n. We allow $\Delta(n; u)$ to count not just divisors in the relevant interval but fractions of the form d^2/d' where d and d' divide n, suitably weighted. For $\lambda \leqslant \frac{1}{3}$, we have

$$\frac{1}{2\pi}\int_0^{2\pi} |\lambda e^{-i\theta} + 1 + e^{i\theta} + \lambda e^{2i\theta}| \, d\theta = \frac{4}{\pi}\left(1 - \frac{\lambda}{3}\right)$$

which is *less* than $H_2 = 4/\pi$ if $\lambda > 0$. This idea led to an improvement in the then best upper bound for $\alpha(2, 1)$.

For larger values of y (6.20) is very useful. We work out the case $r = 2$, when the integral can be estimated fairly precisely because $|1 + e^{i\theta_1 t}|$ is a periodic function of t. In fact the integral w.r.t. t is

$$\begin{cases} \dfrac{4}{\pi}\log L + \left(2 - \dfrac{4}{\pi}\right)\log\dfrac{1}{|\theta_1|} + O(1) & (1/L < |\theta_1| \leqslant 1), \\ 2\log L + O(1). & (|\theta_1| \leqslant 1/L). \end{cases}$$

Hence

$$S_2(x, y) \ll_y x(\log x)^{2y-2} + x(\log x)^{(4/\pi)y - 1}\int_{1/L}^{1} \theta^{-(2-4/\pi)y} \, d\theta.$$

The reader can now see the important rôle played by the parameter y: if $y < 1/(2 - H_2)$, we obtain $S_2(x, y) \ll x(\log x)^{H_2 y - 1}$, an unsharp bound which includes Hooley's original upper estimate for $S_2(x)$. For certain y, it can be improved by the 'λ-method' described in the last paragraph. If $y > 1/(2 - H_2)$ we obtain the sharp bound $S_2(x, y) \ll_y x(\log x)^{2y-2}$ (cf. (6.8)); when $y = 1/(2 - H_2)$ there is an extra factor $\log\log x$. Thus we have obtained the following.

Theorem 66. $\Lambda_2^+ \leqslant 1/(2 - H_2) = 1.375\,96\ldots$, *more precisely for* $y \geqslant 1/(2 - H_2)$ *we have*

$$S_2(x, y) \ll_y x \log^{2y-2} x (\log_2 x)^{\delta((2-H_2)y,\, 1)} . \tag{6.25}$$

6.8 Fourier transforms – an upper bound for Λ_r^+

Theorem 67. *For $r \geqslant 3$ we have $\Lambda_r^+ \leqslant 1 + r^{-1/2}$. More precisely, for $y \geqslant 1 + r^{-1/2}$,*

$$S_r(x, y) \ll_{r,y} x \log^{ry-r} x (\log_2 x)^{\delta(y, 1+r^{-1/2})}. \tag{6.26}$$

Slightly better numerical bounds are available in the range $4 \leqslant r \leqslant 8$ from Hall (1986). These are

$$\Lambda_4^+ < 1.469\,80, \quad \Lambda_5^+ < 1.403\,09, \quad \Lambda_6^+ < 1.373\,50, \quad \Lambda_7^+ < 1.357\,48,$$
$$\Lambda_8^+ < 1.347\,72.$$

An open problem of some interest is to reduce the upper bound $1 + 1/\sqrt{3}$ for Λ_3^+.

In each of Theorems 64, 66, 67 we have an inequality $\Lambda_r^- \geqslant w$, $\Lambda_r^+ \leqslant z$ say, and the proofs also yield useful, if unsharp, results on the boundaries, respectively $S_r(x, w) \ll x \log^{w-1} x (\log_2 x)^A$, $S_r(x, z) \ll x \log^{rz-r} x (\log_2 x)^B$ for suitable A, B. However this is a feature of these particular proofs: there is no implication that such estimates involving just an additional power of $\log_2 x$ hold if w, z be replaced by Λ_r^-, Λ_r^+.

The following proof is based on the argument to be found on p. 124 of Hooley (1979).

Proof of Theorem 67. We write

$$G(t; \theta_1, \theta_2, \ldots, \theta_{r-1}) := \int_1^t |1 + e^{i\theta_1 w} + e^{i\theta_2 w} + \cdots + e^{i\theta_{r-1} w}| \, dw$$

so that the innermost integral in (6.20) is

$$\int_1^L G(t) \frac{dt}{t^2} + O(r);$$

indeed

$$S_r(x, y) \ll_{r,y} \frac{x}{L} \int_{-1}^1 \cdots \int_{-1}^1 \exp\left\{ y \int_1^L G(t) \frac{dt}{t^2} \right\} d\theta_1 \ldots d\theta_{r-1}. \tag{6.27}$$

The Cauchy–Schwarz inequality applied to the integral defining G yields

$$G(t)^2 \leqslant t \int_1^t \left\{ r + \sum_{j<r} 2 \cos w\theta_j + \sum_{0<k<l<r} 2 \cos w(\theta_l - \theta_k) \right\} dw$$
$$\leqslant rt^2 + 2t \sum_{j<r} \min\left(t, \frac{2}{|\theta_j|}\right) + 2t \sum_{0<k<l<r} \min\left(t, \frac{2}{|\theta_l - \theta_k|}\right).$$

The right-hand side is largest when the θ_i have the same sign. Because of this, and the symmetry of the integral (6.27), we may restrict our attention to the range in which $0 \leqslant \theta_1 \leqslant \theta_2 \leqslant \cdots \leqslant \theta_{r-1} \leqslant 1$, and we put $\theta_1 = x_1$,

$\theta_2-\theta_1=x_2,\ldots,\theta_{r-1}-\theta_{r-2}=x_{r-1}$ so that the $x_i\in[0,1]$. We employ the simple lower bounds

$$\theta_j\geqslant\max\{x_m:1\leqslant m\leqslant j\},\quad \theta_l-\theta_k\geqslant\max\{x_m:k<m\leqslant l\};$$

since we have to integrate we may assume that the x_i are distinct so that m above is uniquely determined. Of course the integral (6.27) is not symmetric in these new variables.

We split the range of integration into $(r-1)!$ regions, in each of which there is a permutation ρ on $r-1$ symbols such that $x_{\rho(1)}<x_{\rho(2)}<\cdots<x_{\rho(r-1)}$, and for each m and ρ we denote by $v(m,\rho)$ the number of times $x_{\rho(m)}$ is chosen as maximum above. In the region associated with the permutation ρ we then have

$$G(t)^2\leqslant rt^2+2t\sum_{m=1}^{r-1}v(m,\rho)\min(t,2/x_{\rho(m)}),$$

and so, if $x_{\rho(a)}<1/t\leqslant x_{\rho(a+1)}$ (we put $a=0$ when $1/t\leqslant x_{\rho(1)}$, $a=r-1$ when $x_{\rho(r-1)}<1/t$), we have

$$G(t)\leqslant R(a)t+2\sum_{m=a+1}^{r-1}(tv(m,\rho)/x_{\rho(m)})^{1/2} \tag{6.28}$$

where

$$R(a):=\left(r+2\sum_{m=1}^{a}v(m,\rho)\right)^{1/2}. \tag{6.29}$$

Now let b be such that $x_{\rho(b)}<1/L\leqslant x_{\rho(b+1)}$ (with $b=0$ or $r-1$ if $1/L\leqslant x_{\rho(1)}$ or $x_{\rho(r-1)}<1/L$). Then

$$\begin{aligned}\int_1^L G(t)\frac{dt}{t^2}\leqslant&\int_1^{1/x_{\rho(r-1)}}\frac{r\,dt}{t}\\&+\int_{1/x_{\rho(r-1)}}^{1/x_{\rho(r-2)}}\left(R(r-2)+2\left(\frac{v(r-1,\rho)}{tx_{\rho(r-1)}}\right)^{1/2}\right)\frac{dt}{t}\\&+\cdots+\int_{1/x_{\rho(b+1)}}^{L}\left(R(b)+2\sum_{b<m<r}\left(\frac{v(m,\rho)}{tx_{\rho(m)}}\right)^{1/2}\right)\frac{dt}{t}.\end{aligned}$$

The integrals involving the factor $t^{-3/2}$ on the right-hand side are either $\ll v(m,\rho)^{1/2}$ $(b+1<m<r)$, or

$$\ll v(m,\rho)^{1/2}\frac{x_{\rho(b+1)}^{1/2}}{x_{\rho(m)}^{1/2}}\quad(b<m<r),$$

so that their total contribution is $\ll_r 1$. Thus

$$\begin{aligned}\int_1^L G(t)\frac{dt}{t^2}\leqslant& R(b)\log L+(R(b+1)-R(b))\log\frac{1}{x_{\rho(b+1)}}\\&+\cdots+(r-R(r-2))\log\frac{1}{x_{\rho(r-1)}}+O_r(1).\end{aligned} \tag{6.30}$$

(The right-hand side is simply $r\log L+O_r(1)$ if $b=r-1$).

Lemma 67.1. *Let $b < r - 1$, and $z_{b+1}, z_{b+2}, \ldots, z_{r-1}$ be such that $z_{r-1} \geqslant 1$, $z_{r-2} + z_{r-1} \geqslant 2, \ldots, z_{b+1} + z_{b+2} + \cdots + z_{r-1} \geqslant r - 1 - b$, with exactly σ cases of equality. Then*

$$\int_{1/L}^{1} \frac{du_{b+1}}{u_{b+1}^{z_{b+1}}} \int_{u_{b+1}}^{1} \frac{du_{b+2}}{u_{b+2}^{z_{b+2}}} \cdots \int_{u_{r-2}}^{1} \frac{du_{r-1}}{u_{r-1}^{z_{r-1}}} \ll L^{z-r+b+1} (\log L)^{\sigma}$$

where $z = z_{b+1} + z_{b+2} + \cdots + z_{r-1}$, and the implied constant depends only on the z_i.

Proof. This is by induction on r. The result holds for $r = b + 2$ and we suppose it holds for $r - 1$, where $r \geqslant b + 3$. We have

$$\int_{u_{r-2}}^{1} \frac{du_{r-1}}{u_{r-1}^{z_{r-1}}} \ll \frac{(\log L)^{\delta(z_{r-1}, 1)}}{u_{r-2}^{z_{r-1}-1}}$$

because $z_{r-1} \geqslant 1$, $u_{r-2} \geqslant 1/L$. Hence we may reduce to an $(r-b-2)$-fold integral with exponents $z_{b+1}, z_{b+2}, \ldots, z_{r-3}, z'_{r-2}$ where $z'_{r-2} = z_{r-2} + z_{r-1} - 1$. For $b < m \leqslant r - 2$, we have $z_m + z_{m+1} + \cdots + z'_{r-2} \geqslant r - m - 1$ as required, and there are $\sigma' = \sigma - \delta(z_{r-1}, 1)$ cases of equality. By the induction hypothesis, we obtain altogether

$$\ll (\log L)^{\delta(z_{r-1}, 1)} L^{z' - (r-1) + b + 1} (\log L)^{\sigma'}$$

where $z' = z_{b+1} + z_{b+2} + \cdots + z'_{r-2} = z - 1$, and this is all we need.

Now we consider the integral (6.27), confining our attention to the region associated with the permutation ρ, and the range where $x_{\rho(b)} < 1/L \leqslant x_{\rho(b+1)}$. If $b = r - 1$, we have $x_m \leqslant 1/L$ for every m so that the volume to be integrated over is $\leqslant L^{1-r}$. So we obtain

$$\ll \frac{x}{\log x} L^{1-r} \exp(ry \log L) \ll x \log^{ry-r} x$$

using (6.30). Now let $b < r - 1$. We multiply (6.30) by y, exponentiate, and insert in (6.27). We apply Lemma 67.1, with $u_i = x_{\rho(i)}$, $z_i = y(R(i) - R(i-1))$, $b + 1 \leqslant i < r$. We must establish the inequalities

$$\left.\begin{aligned} (r - R(r-2))y &\geqslant 1, \\ (r - R(r-3))y &\geqslant 2, \\ &\cdots \\ z = (r - R(b))y &\geqslant r - 1 - b, \end{aligned}\right\} \tag{6.31}$$

and determine how often equality occurs. To this end we maximize $R(s)$ as a function of ρ.

Let us write $s = r - 1 - q$ $(q \geqslant 1)$. We have to maximize $v(1, \rho) + v(2, \rho) + \cdots + v(s, \rho)$, which is the number of times one or other of $x_{\rho(1)}$, $x_{\rho(2)}, \ldots, x_{\rho(s)}$ is the largest x_i in the sum $x_{k+1} + x_{k+2} + \cdots + x_l$ $(0 \leqslant k < l < r)$. Since these are the smallest variables, equivalently *none* of $x_{\rho(r-q)}$, $x_{\rho(r-q+1)}, \ldots, x_{\rho(r-1)}$ occur in this sum. We write the numbers $\rho(r-q), \ldots,$

$\rho(r-1)$ in increasing order, as $m_1, m_2, \ldots, m_q$ say, and for convenience in what follows we introduce $m_0 = 0$, $m_{q+1} = r$. If, for all i, $m_i \notin (k, l]$, then there exists u $(0 \leqslant u \leqslant q)$ such that $m_u \leqslant k < l < m_{u+1}$, which gives in total

$$\sum_{u=0}^{q} \binom{m_{u+1} - m_u}{2} = \frac{1}{2} \sum_{u=0}^{q} (m_{u+1} - m_u)^2 - \tfrac{1}{2} r$$

choices for k and l. Thus

$$\sum_{i=1}^{s} v(i, \rho) = \frac{1}{2} \sum_{u=0}^{q} (m_{u+1} - m_u)^2 - \tfrac{1}{2} r$$
$$\leqslant \binom{r-q}{2}, \tag{6.32}$$

observing that the *positive* integers $m_{u+1} - m_u$ have sum r: the sum on the right of (6.32) is maximal when $m_{u+1} - m_u$ takes the value $r - q$ once and equals 1 else. From (6.29) and (6.32), we deduce that (since $r - q = s + 1$),

$$R(s) \leqslant \sqrt{(r + s(s+1))} \quad (0 \leqslant s < r). \tag{6.33}$$

We show that, when $y \geqslant 1 + r^{-1/2}$, the inequalities (6.31) hold, whatever the value of b, that is

$$(r - R(s))y \geqslant r - 1 - s \quad (0 \leqslant s < r - 1).$$

In view of (6.33) we just need

$$(r + s)y^2 - 2ry + r - s - 1 \geqslant 0$$

and this holds: the left-hand side is at least $r(y-1)^2 - 1 \geqslant 0$. Moreover equality requires both $y = 1 + r^{-1/2}$ and $s = 0$.

That part of the integral (6.27) under consideration is

$$\ll xL^{R(b)y - b - 1} \int_{1/L}^{1} \frac{\mathrm{d}x_{\rho(b+1)}}{x_{\rho(b+1)}^{(R(b+1) - R(b))y}} \cdots \int_{x_{\rho(r-2)}}^{1} \frac{\mathrm{d}x_{\rho(r-1)}}{x_{\rho(r-1)}^{(r - R(r-2))y}},$$

and we have shown that the conditions of Lemma 67.1 are satisfied. Since $z = (r - R(b))y$, the above is

$$\ll xL^{ry - r}(\log L)^{\sigma}$$

where $\sigma \leqslant 1$, with equality if and only if $y = 1 + r^{-1/2}$, $b = 0$. We put all these integrals together to obtain (6.26). This completes the proof.

The exact order of magnitude of the right-hand side of (6.20) has not been determined in any of the cases $r \geqslant 3$, $y \leqslant 1 + r^{-1/2}$. Let us denote by $\beta(r, y)$ the infimum of the numbers ξ for which it is $\ll L^{\xi}$. We must have $\beta(r, y) \geqslant \max(H_r y - 1, ry - r)$ and we see that there is equality when $r = 2$. We prove in the notes that $\beta(r, y) \leqslant \max(y\sqrt{r} - 1, ry - r)$. The simplest hypothesis is that the exact order of magnitude is

$$L^{\max(H_r y - 1, ry - r)}(\log L)^{\delta(y, (r-1)/(r - H_r))}.$$

Notes on Chapter 6

The original definition of Λ_r^-, Λ_r^+ which appeared in Hall (1986) was different. When that paper was written the function $\alpha(r, y)$ had not been evaluated for every y. In view of Hooley's opinion (Hooley (1979), p. 116) that $\alpha(2, 1) > 0$ it was not at all clear that the graph consisted of two straight lines – $\alpha(r, y)$ might have been differentiable. Hall defined

$$\Lambda_r^- = \sup\{y : \alpha(r, z) = z - 1 \quad (0 < z \leqslant y)\},$$

$$\Lambda_r^+ = \inf\{y : \alpha(r, z) = rz - r \quad (y \leqslant z < \infty)\}$$

(which are of course both equal to 1). The stronger definition came later, the results in Hall (1986) holding good as stated.

Hooley (1979) used the Fourier transform method to estimate $S_r(x, 1)$ from above, and so it pre-dates the iteration inequalities of Hall & Tenenbaum (1982, 1984) which were used to show that $\alpha(2, 1) < H_2 - 1$, $\alpha(3, 1) < H_3 - 1$, $\alpha(4, 1) < 1$, and (more significantly now), to evaluate $\alpha(r, y)$, $y \leqslant \frac{1}{2}$. Hooley stated (Theorem 1B) a result equivalent to $S_r(x, 1) \ll x(\log x)^{\sqrt{r} - 1}$, proving this when $r = 3$, and the mistaken conviction of Hall & Tenenbaum (1984) that the proof could not be extended beyond $r = 4$ has led to some confusion. To set the record straight, we show now that the method can be made to yield $S_r(x, y) \ll_{r,y} x(\log x)^{y\sqrt{r} - 1}$ for $y < 1 + r^{-1/2}$ and every r.

We put $1/L = \sigma$, for convenience, and notice that (6.30) may be re-written in a form which avoids b, viz.

$$\int_1^L G(t)\frac{dt}{t^2} \leqslant r^{1/2}\log L + (R(1) - R(0))\log\frac{1}{\sigma + x_{\rho(1)}} + \cdots + (r - R(r-2))\log\frac{1}{\sigma + x_{\rho(r-1)}} + O_r(1).$$

The contribution to (6.27) is

$$\ll xL^{y\sqrt{r} - 1}\int_0^1 \frac{dx_{\rho(1)}}{(\sigma + x_{\rho(1)})^{z_1}} \cdots \int_{x_{\rho(r-2)}}^1 \frac{dx_{\rho(r-1)}}{(\sigma + x_{\rho(r-1)})^{z_{r-1}}}$$

where, as before, $z_i = (R(i) - R(i-1))y$. We claim that for $y < 1 + r^{-1/2}$ we have $z_1 + z_2 + \cdots + z_s < s$ $(1 \leqslant s < r)$, and that this latter condition implies that the multiple integral above is bounded as $\sigma \to 0$. A similar assertion (with $z_1 + z_2 + \cdots + z_s \leqslant s$, and the integral $\ll \log^{r-1}(1/\sigma)$) is stated and proved in Hall (1986): the stronger hypothesis and conclusion may be treated in the same way. The former part of our claim is that $(R(s) - \sqrt{r})y < s$, or, from (6.33), that $(\sqrt{(r + s(s+1))} - \sqrt{r})(1 + 1/\sqrt{r}) \leqslant s$. This last inequality is valid for $s = 0$, $s = r - 1$; and we just observe that the left-hand side is a convex function of s. This gives the result stated: a corollary is $\beta(r, y) \leqslant \max(y\sqrt{r} - 1, ry - r)$.

Exercises on Chapter 6

60. Prove that

$$\Delta_r(2^\alpha) = \binom{r-1}{0} + \binom{r-1}{1} + \cdots + \binom{r-1}{\alpha}.$$

61. Prove that, for $y, z > 0$,

$$Z_r(x, y)^2 \leqslant Z_{r-1}(x, y^2/z) Z_r(x, z) \sum_{\substack{dd' \leqslant x \\ |\log(d/d')| < 1}} \frac{\mu^2(dd')}{dd'} z^{\omega(dd')}$$

and hence that

$$2\alpha(r, y) \leqslant \alpha(r, z) + (2z - 1)^+ + \alpha(r - 1, y^2/z).$$

(This is (6.19) with $s = 2$). Put $z = \frac{1}{2}$ and deduce that

$$\alpha(r, y) \leqslant 2^{-r}((2y)^{2^{r-1}} - 1) - \tfrac{1}{2} \quad (y \geqslant \tfrac{1}{2}).$$

Now substitute $\alpha(2, z) \leqslant z^2 - \frac{3}{4}$ in the iteration inequality above and estimate $\alpha(r, 1)$ by choosing z optimally.

62. Prove that for $k \in \mathbb{Z}^+$ the moment

$$M_r^{(2k)} := \frac{1}{(2\pi)^{r-1}} \int_0^{2\pi} \cdots \int_0^{2\pi} |1 + e^{i\varphi_1} + \cdots + e^{i\varphi_{r-1}}|^{2k} \, d\varphi_1 \ldots d\varphi_{r-1}$$

is a polynomial in r of degree k, with leading term $k!r^k$. Evaluate $M_r^{(2)}$ and $M_r^{(4)}$, and deduce (by Hölder) that $r^{3/2}/(2r^2 - r)^{1/2} \leqslant H_r \leqslant \sqrt{r}$.

7

Hooley's Δ_r-functions – the critical interval

7.1 Introduction

We continue our study of Hooley's function $\Delta_r(n)$, defined by (6.2), and we recall Theorem 60 which states that

$$S_r(x, y) := \sum_{n \leqslant x} \Delta_r(n) y^{\omega(n)}$$
$$\gg_{r,y} x(\log x)^{y-1+(r-1)(y-1)^+}(\log_2 x)^{\delta(y,1)}.$$

The simplest hypothesis is that the right-hand side represents the exact order of magnitude of $S_r(x, y)$.

Conjecture. *For each $r \geqslant 2$ and $y > 0$ there exists a constant $A(r, y)$ such that*

$$S_r(x, y) \sim A(r, y)x(\log x)^{y-1+(r-1)(y-1)^+}(\log_2 x)^{\delta(y,1)} \tag{7.1}$$

This is quite strong, and we review what we know so far. There is no question about the order of magnitude as a lower bound; moreover for each r there exists a 'critical interval' $[\Lambda_r^-, \Lambda_r^+]$ outside of which (7.1) actually holds if we replace $\sim$ by $\asymp$. Moreover we know that $\frac{1}{2} \leqslant \Lambda_r^- \leqslant 1 \leqslant \Lambda_r^+ \leqslant 1 + r^{-1/2}$. We do not possess an asymptotic formula for any combination of r and y – but the techniques of the previous chapter could be made to yield explicit constants in the upper and lower bounds which could be computed and perhaps sharpened.

Far more serious is the critical interval. At the time of writing, the best information we have about this is set out in Theorems 64 and 67. The methods of Chapter 6 extend into the interval to some extent: we have the iteration inequality (6.19), the convexity of the function $\alpha(r, y)$, and special devices such as the λ-method to hand, and yet the correct exponent of $\log x$ in the upper bound for $S_r(x, y)$ eludes us. The object of the present chapter is to describe a completely different method which evaluates $\alpha(r, y)$ for every r and y, that is, determines $S_r(x, y)$ to within a factor $(\log x)^{o(1)}$. Since this method does not (and seems unlikely to) yield sharp bounds, it is only employed in the critical interval. We obtain the following results.

Theorem 70. *For each $r \geqslant 2$ we have that*

$$S_r(x, 1) \ll_r x \exp\left\{\left(r - 1 + \frac{30}{\log_3 x}\right)\sqrt{(r \log_2 x \cdot \log_3 x)}\right\}, \tag{7.2}$$

and, for $y > 1$,

$$S_r(x, y) \ll_{r,y} x(\log x)^{ry-r} \exp\left\{(r-1)\sqrt{\left(2r \log_2 x \cdot \log \frac{5y}{y-1}\right)}\right\}.$$

Theorem 71. *For each $r \geqslant 2$ and $0 < y < 1$ we have that*

$$S_r(x, y) \ll_{r,y} x(\log x)^{y-1} \exp\left\{\frac{4 \log(r+1)}{(1-y)}(\log_3 x)^2\right\}. \tag{7.3}$$

An immediate corollary is the formula (6.6) for $\alpha(r, y)$. The proofs of these results rest on a common foundation, the fundamental lemma of §7.2. However, the method has two variants in which the lemma is used in ways which at first sight appear to differ only in a matter of technical preference, but emerge as much more divergent. In the proof of each theorem, the argument breaks down if y is not in the correct range.

This is not yet perfectly understood, and the reader who has studied both proofs is encouraged to consider the matter for himself. To complicate it further, we mention that there is in fact a 'unified' treatment of the whole range $0 < y < \infty$, resembling the first variant presented here, but giving an inferior result to (7.3) when $0 < y < 1$ and, apparently, requiring the introduction of an extra parameter. (See the Notes.)

7.2 Notation. The fundamental lemma

We recall from (6.1) the definition

$$\Delta(n; u_1, u_2, \ldots, u_{r-1}) = \operatorname{card}\{d_1 d_2 \ldots d_{r-1} | n : u_i < \log d_i \leqslant u_i + 1 \quad (1 \leqslant i < r)\}$$

and we set (suppressing the variable r on the left)

$$M_q(n) := \int_{-\infty}^{\infty} \cdots \int_{-\infty}^{\infty} \Delta(n; u_1, u_2, \ldots, u_{r-1})^q \, du_1 \, du_2 \ldots du_{r-1} \tag{7.4}$$

for $q \geqslant 1$. Of course the integrand is supported on a finite range and the integral is finite. We expect that, for large q, the integral will be dominated by the largest values of the integrand, in fact

$$\lim_{q \to \infty} M_q(n)^{1/q} = \Delta_r(n) \tag{7.5}$$

(cf. Exercise 74). We do not use this; we prove instead the following proposition which does not include (7.5) but is similar in content, and in an applicable form.

Theorem 72. *For each $r \geqslant 2$ and $q \geqslant 1$ we have*

$$M_1(n) = \tau_r(n), \tag{7.6}$$

$$M_q(n) \leqslant \Delta_r(n)^{q-1} M_1(n) \leqslant \tau_r(n)^q, \tag{7.7}$$

$$\Delta_r(n)^q \leqslant 2^{(r-1)(q-1)} M_q(n). \tag{7.8}$$

We prove (7.8), the rest being clear. Let $\Delta_r(n) = \Delta(n; v_1, v_2, \dots, v_{r-1})$ say, and $v_i \leqslant u_i < v_i + 1$, $\varepsilon_i = 0$ or 1, for $1 \leqslant i < r$. Then

$$\Delta(n; v_1, v_2, \dots, v_{r-1}) \leqslant \sum \Delta(n; u_1 - \varepsilon_1, u_2 - \varepsilon_2, \dots, u_{r-1} - \varepsilon_{r-1})$$

where the sum is over all 2^{r-1} possible choices of the ε_i. We deduce that

$$\Delta_r(n)^q \leqslant 2^{(r-1)(q-1)} \sum \Delta(n; u_1 - \varepsilon_1, u_2 - \varepsilon_2, \dots, u_{r-1} - \varepsilon_{r-1})^q$$

and we integrate this over the cube $v_i \leqslant u_i < v_i + 1$ $(1 \leqslant i < r)$. We obtain (7.8) because there is no overlapping of the ranges of integration for different choices of $(\varepsilon_1, \varepsilon_2, \dots, \varepsilon_{r-1})$.

Definition. *For non-negative $a_0, a_1, \dots, a_{r-1}$ and real w, set*

$$\begin{aligned} N(n; w; a_0, a_1, \dots, a_{r-1}) &= \int_{-\infty}^{\infty} \cdots \int_{-\infty}^{\infty} \Delta(n; u_1, u_2, \dots, u_{r-1})^{a_0} \\ &\quad \times \Delta(n; u_1 - w, u_2, \dots, u_{r-1})^{a_1} \\ &\quad \times \Delta(n; u_1, u_2 - w, u_3, \dots, u_{r-1})^{a_2} \cdots \\ &\quad \times \Delta(n; u_1, \dots, u_{r-1} - w)^{a_{r-1}} \, du_1 \dots du_{r-1} \end{aligned} \tag{7.9}$$

where w is subtracted in turn from each u_i in the last $r-1$ factors.

From Hölder's inequality, we always have

$$N(n; w; a_0, a_1, \dots, a_{r-1}) \leqslant M_{a_0 + a_1 + \cdots + a_{r-1}}(n).$$

Definition. *For $s \geqslant 1$, set*

$$D_s(m) := \sum_{d \mid m} \Delta_s(d) \tag{7.10}$$

(so that for example $D_1(m) = \tau(m)$). We can now state the result which lies at the centre of the method presented in this chapter in any of its forms. *From now on q, and the exponents a_i, are restricted to integer values.*

Theorem 73 (Fundamental Lemma). *Let q be an integer $\geqslant 2$, and*

$a_0, a_1, \ldots, a_{r-1}$ be integers such that $0 \leqslant a_i < q$ $(0 \leqslant i < r)$ and $a_0 + a_1 + \cdots + a_{r-1} = q$. Then

$$\sum_p N(n; \log p; a_0, a_1, \ldots, a_{r-1}) \frac{\log p}{p} \leqslant C \cdot 2^{q-1} D_{r-1}(n) \tau_r(n)^{1/(q-1)} M_q(n)^{(q-2)/(q-1)}, \tag{7.11}$$

where the sum is over all primes p and C is an absolute constant, precisely

$$C := \sup_z \sum_{z \leqslant p < ez} \frac{\log p}{p}.$$

It may be helpful at this point to write down a simple upper bound for the left-hand side of (7.11) for comparison. To do this we first show that the sum over p is really over a finite range. Now if $\Delta(n; u_1, u_2, \ldots, u_{r-1}) \neq 0$ then $-1 \leqslant u_i < \log n$ for every i. Next, there must be at least two non-zero exponents a_i, (because each is less than their sum), and so each u_i appears at least once in the integrand in (7.9) without w subtracted. Hence the ranges of integration in (7.9) could as well be restricted to $[-1, \log n]$. If n is to have any divisor such that $u_i - w < \log d \leqslant u_i + 1 - w$ we also see that we must have $u_i - \log n < w \leqslant u_i + 1$, that is $|w| \leqslant \log n + 1$, and this condition actually applies because $a_1, a_2, \ldots, a_{r-1}$ are not all zero. So the sum over p in (7.11) could just as well be restricted to the range $2 \leqslant p < en$, and then $\ll M_q(n) \log n$ is a valid upper bound. That given is not always as good: it is, however, what we want.

Proof of Theorem 73. The first step is to show that it is sufficient to prove (7.11) when just two of the a_i are non-zero. Suppose then that there are at least three, namely $a_{i_0}, a_{i_1}, \ldots, a_{i_s}$. Let $a_{i_1} + a_{i_2} + \cdots + a_{i_s} = a$, and for $1 \leqslant t \leqslant s$ set $k_t = a/a_{i_t}$ so that $\sum k_t^{-1} = 1$. We apply Hölder's inequality, with exponents k_t, to show that the left-hand side of (7.11) does not exceed

$$\prod_{t=1}^{s} \left\{ \sum_p N(n; \log p; 0, 0, \ldots, a_{i_0}, \ldots, a, \ldots, 0) \frac{\log p}{p} \right\}^{1/k_t}$$

and it will be sufficient to establish the desired inequality for each of these inner sums.

There are now two cases to consider (unless $r = 2$) according as $i_0 = 0$ or not. We begin with the first case, $i_0 = 0$: we may suppose that $i_1 = 1$. We have $a_0 = q - a$, $a_1 = a$, $a_i = 0$ $(2 \leqslant i < r)$. Also $1 \leqslant a \leqslant q - 1$. We seek an upper bound for the sum

$$\sum_p \Delta(n; u_1 - \log p, u_2, \ldots, u_{r-1})^a \frac{\log p}{p} = \sum'_{d_1^{(1)} \cdots d_{r-1}^{(1)} | n} \sum'_{d_1^{(2)} \cdots d_{r-1}^{(2)} | n} \cdots \sum'_{d_1^{(a)} \cdots d_{r-1}^{(a)} | n} \sum'_p \frac{\log p}{p} \tag{7.12}$$

where the dash denotes the following conditions of summation:

$$\left.\begin{array}{l} u_1 - \log p < \log d_1^{(k)} \leqslant u_1 - \log p + 1 \quad (1 \leqslant k \leqslant a), \\ u_i < \log d_i^{(k)} \leqslant u_i + 1 \quad (1 \leqslant k \leqslant a, 2 \leqslant i < r). \end{array}\right\} \tag{7.13}$$

Therefore

$$\left.\begin{array}{r} \max_k \log d_1^{(k)} - \min_k \log d_1^{(k)} < 1, \\ u_1 - \min_k \log d_1^{(k)} \leqslant \log p < u_1 + 1 - \max_k \log d_1^{(k)}, \end{array}\right\} \tag{7.14}$$

whence, from the second part of (7.14), we always have in (7.12)

$$\sum{}' \frac{\log p}{p} \leqslant C.$$

Hence the sum (7.12) does not exceed

$$C \sum_{d_1^{(1)} \cdots d_{r-1}^{(1)} \mid n}'' \; \sum_{d_1^{(2)} \cdots d_{r-1}^{(2)} \mid n}'' \cdots \sum_{d_1^{(a)} \cdots d_{r-1}^{(a)} \mid n}'' 1 \tag{7.15}$$

where the double dash means that the summation conditions are now the second line of (7.13) and the first line of (7.14).

Definition. *For* $a \in \mathbb{Z}^+$, $u_2, u_3, \ldots, u_{r-1} \in \mathbb{R}$,

$$V_a(n; u_2, \ldots, u_{r-1}) := \int_{-\infty}^{\infty} \Delta(n; u_1, u_2, \ldots, u_{r-1})^a \, \mathrm{d}u_1. \tag{7.16}$$

Consider the integral

$$\int_{-\infty}^{\infty} \{\Delta(n; u_1 - 1, u_2, \ldots, u_{r-1}) + \Delta(n; u_1, u_2, \ldots, u_{r-1})\}^a \, \mathrm{d}u_1$$

which, firstly, does not exceed $2^a V_a(n; u_2, \ldots, u_{r-1})$. But, secondly, this integral may be written as

$$\sum_{d_1^{(1)} \cdots d_{r-1}^{(1)} \mid n}^{*} \cdots \sum_{d_1^{(a)} \cdots d_{r-1}^{(a)} \mid n}^{*} \left(2 - \log\left\{\max_k d_1^{(k)} / \min_k d_1^{(k)}\right\}\right)^+ \tag{7.17}$$

where the * means that the summation condition is simply the second line of (7.13). If we compare (7.15) with (7.17), noting that in (7.15) the first line of (7.14) is an extra summation condition, we may deduce that the sum (7.12) does not exceed $C \cdot 2^a V_a(n; u_2, \ldots, u_{r-1})$, and hence that

$$\begin{aligned} &\sum_p N(n; \log p; q-a, a, 0, \ldots, 0) \frac{\log p}{p} \\ &\leqslant C \cdot 2^a \int_{-\infty}^{\infty} \cdots \int_{-\infty}^{\infty} \Delta(n; u_1, u_2, \ldots, u_{r-1})^{q-a} \\ &\qquad \times V_a(n; u_2, \ldots, u_{r-1}) \, \mathrm{d}u_1 \, \mathrm{d}u_2 \ldots \mathrm{d}u_{r-1} \end{aligned}$$

$$\leqslant C\cdot 2^a \int_{-\infty}^{\infty} \cdots \int_{-\infty}^{\infty} V_{q-a}(n; u_2, \ldots, u_{r-1})$$
$$\times V_a(n; u_2, \ldots, u_{r-1})\, du_2 \ldots du_{r-1}. \tag{7.18}$$

Recall that $a, q-a \in [1, q-1]$. Hölder's inequality gives

$$V_a(n; u_2, \ldots, u_{r-1}) \leqslant V_1(n; u_2, \ldots, u_{r-1})^{(q-a)/(q-1)}$$
$$\times V_q(n; u_2, \ldots, u_{r-1})^{(a-1)/(q-1)},$$

together with a corresponding inequality for V_{q-a}. Therefore the last integral in (7.18) does not exceed

$$C\cdot 2^{q-1} \int_{-\infty}^{\infty} \cdots \int_{-\infty}^{\infty} V_1(n; u_2, \ldots, u_{r-1})^{q/(q-1)}$$
$$\times V_q(n; u_2, \ldots, u_{r-1})^{(q-2)/(q-1)}\, du_2 \ldots du_{r-1}$$

and we notice that

$$V_1(n; u_2, \ldots, u_{r-1}) = \sum_{d_1 | n} \Delta\left(\frac{n}{d_1}; u_2, \ldots, u_{r-1}\right) \leqslant D_{r-1}(n)$$

where D was defined in (7.10). We apply this last inequality to the factor V_1 in the integrand above, removing $D_{r-1}(n)$ as a constant factor and leaving behind $V_1^{1/(q-1)}$. Finally, we apply Hölder's inequality (again!) with exponents $q-1$, $(q-1)/(q-2)$. We have

$$\int_{-\infty}^{\infty} \cdots \int_{-\infty}^{\infty} V_1(n; u_2, \ldots, u_{r-1})\, du_2 \ldots du_{r-1} = M_1(n) = \tau_r(n),$$

$$\int_{-\infty}^{\infty} \cdots \int_{-\infty}^{\infty} V_q(n; u_2, \ldots, u_{r-1})\, du_2 \ldots du_{r-1} = M_q(n),$$

and we obtain the inequality (7.11) for the special sum (7.18). This completes our treatment of the first of our two cases.

It remains to deal with the second case, in which $i_0 > 0$: we may suppose $i_0 = 1$, $i_1 = 2$. First, we reduce to the case where the only non-zero exponents a_1 and a_2 are 1 and $q-1$. Suppose $\min(a_1, a_2) > 1$ (so that $q \geqslant 4$). We write $N(n; w; 0, a_1, a_2, 0, 0, \ldots, 0)$ in the form

$$\int_{-\infty}^{\infty} \cdots \int_{-\infty}^{\infty} \Delta(n; u_1 - w, \ldots, u_{r-1})^{b_1} \Delta(n; u_1, u_2 - w, \ldots, u_{r-1})^{a_2 - b_2}$$
$$\times \Delta(n; u_1 - w, \ldots, u_{r-1})^{a_1 - b_1} \Delta(n; u_1, u_2 - w, \ldots, u_{r-1})^{b_2}\, du_1 \ldots du_{r-1}$$

and solve the equations $k_1 b_1 = 1$, $k_1(a_2 - b_2) = q-1$, $k_2(a_1 - b_1) = q-1$, $k_2 b_2 = 1$, $k_1^{-1} + k_2^{-1} = 1$, for b_1, b_2, k_1, k_2. We find that $k_1 = (q-2)/(a_2 - 1)$, $k_2 = (q-2)/(a_1 - 1)$ and we apply Hölder's inequality, with these exponents, to the integral above to obtain that

$$N(n; w; 0, a_1, a_2, 0, \ldots, 0)$$
$$\leqslant N(n; w; 0, 1, q-1, 0, \ldots, 0)^{1/k_1} N(n; w; 0, q-1, 1, 0, \ldots, 0)^{1/k_2}. \quad (7.19)$$

We substitute $\log p$ for w, insert (7.19) into (7.11), and apply Hölder again to the sum over p. This gives the required reduction.

We assume now that $a_1 = 1$, $a_2 = q-1$, so that the left-hand side of (7.11) is

$$\sum_p \frac{\log p}{p} \int_{-\infty}^{\infty} \cdots \int_{-\infty}^{\infty} \Delta(n; u_1 - \log p, u_2, \ldots, u_{r-1})$$
$$\times \Delta(n; u_1, u_2 - \log p, u_3, \ldots, u_{r-1})^{q-1} du_1 \ldots du_{r-1} \quad (7.20)$$

and we make the substitution $u_2 = w + \log p$. The second factor in the integrand above is then independent of p, and we therefore seek an upper bound for the sum

$$\sum_p \Delta(n; u_1 - \log p, w + \log p, u_3, \ldots, u_{r-1}) \frac{\log p}{p}$$
$$= \sum_p \sum_{d_1 \cdots d_{r-1} | n}' \frac{\log p}{p} \quad (7.21)$$

where the dash implies the summation conditions

$$u_1 - \log p < \log d_1 \leqslant u_1 - \log p + 1,$$
$$w + \log p < \log d_2 \leqslant w + \log p + 1,$$
$$u_i < \log d_i \leqslant u_i + 1 \quad (3 \leqslant i < r).$$

The first two conditions involve

$$u_1 + w < \log d_1 d_2 \leqslant u_1 + w + 2$$

and so the sum (7.21) does not exceed

$$\sum_{\substack{d_1 d_2 | n \\ u_1 + w \leqslant \log d_1 d_2 < u_1 + w + 2}} \Delta(n/d_1 d_2; u_3, \ldots, u_{r-1}) \sum{}'' \frac{\log p}{p} \quad (7.22)$$

where the double dash means that we require

$$\max(u_1 - \log d_1, w - \log d_2 - 1) \leqslant \log p$$
$$\leqslant \min(u_1 - \log d_1 + 1, w - \log d_2),$$

so that the inner sum in (7.22) does not exceed C. Hence the sum (7.21) is

$$\leqslant C \sum_{d_1 | n} \left\{ \Delta\left(\frac{n}{d_1}; u_1 + w - \log d_1, u_3, \ldots, u_{r-1}\right)\right.$$
$$\left. + \Delta\left(\frac{n}{d_1}; u_1 + w - \log d_1 + 1, u_3, \ldots, u_{r-1}\right)\right\} \leqslant 2CD_{r-1}(n)$$

after the definition (7.10) of D. We insert this last upper bound into (7.20) to deduce that

$$\sum_p N(n; \log p; 0, 1, q-1, 0, \ldots, 0)\frac{\log p}{p}$$

$$\leqslant 2CD_{r-1}(n)\int_{-\infty}^{\infty}\cdots\int_{-\infty}^{\infty}\Delta(n; u_1, w, u_3, \ldots, u_{r-1})^{q-1}\,du_1\,dw\ldots du_{r-1}.$$

The final application of Hölder's inequality may be left to the reader: he must be an expert on this by now.

7.3 First variant – differential inequalities

The method of this chapter begins with the identity, valid for integers n and primes p not dividing n,

$$\begin{aligned}\Delta(np; u_1, u_2, \ldots, u_{r-1}) &= \Delta(n; u_1, u_2, \ldots, u_{r-1})\\ &+ \Delta(n; u_1 - \log p, u_2, \ldots, u_{r-1}) + \cdots + \Delta(n; u_1, \ldots, u_{r-1} - \log p)\end{aligned} \quad (7.23)$$

in which there are r terms on the right, $\log p$ being subtracted in turn from each u_i in the last $r-1$ of these terms. We raise to a high power q and integrate, obtaining $M_q(np)$ on the left and on the right r diagonal terms each equal to $M_q(n)$ (since the shift of any variable by $\log p$ does not affect the integral), and $r^q - r$ off-diagonal terms $N(n; \log p; a_0, a_1, \ldots, a_{r-1})$. The treatment of these off-diagonal terms will involve the fundamental lemma, Theorem 73. Thus the idea is to use high moments to show that the terms on the right of (7.23) usually behave almost independently, so that, for most n and p, $M_q(np)$ is not too large compared with $M_q(n)$. As explained above, there are two variants of the method, just according to how the extra prime p is introduced. However this affects the technical 'working-out' in such a way that one variant yields Theorem 70, the other Theorem 71. In either case we work with squarefree numbers only (which is inherent in the form of (7.23)), finally switching over to all numbers very much as in the previous chapter. In the first variant, we require two lemmas.

Lemma 70.1. *For $\sigma > 1$ we have*

$$\sum_p \frac{\log p}{p^\sigma} < \frac{-\zeta'(\sigma)}{\zeta(\sigma)} < \frac{1}{\sigma - 1}. \quad (7.24)$$

Proof. We have

$$\zeta(\sigma) = \frac{1}{\sigma-1} + \frac{1}{2} + \frac{\sigma(\sigma+1)}{2}\int_1^\infty \frac{\{x\}(1-\{x\})}{x^{\sigma+2}}\,dx$$

where $\{x\}$ denotes the fractional part of x. The integral is positive, and

$$\zeta(\sigma) > \frac{1}{\sigma - 1} + \tfrac{1}{2} \quad (\sigma > 1).$$

Differentiation yields

$$-\zeta'(\sigma) = \frac{1}{(\sigma-1)^2} + \frac{\sigma(\sigma+1)}{2} \int_1^\infty \frac{\{x\}(1-\{x\})\log x}{x^{\sigma+2}}\,dx$$
$$-\frac{(2\sigma+1)}{2} \int_1^\infty \frac{\{x\}(1-\{x\})}{x^{\sigma+2}}\,dx,$$

in which the third term on the right is negative, and the second does not exceed

$$\frac{\sigma(\sigma+1)}{8} \int_1^\infty \frac{\log x}{x^{\sigma+2}}\,dx = \frac{\sigma}{8(\sigma+1)}.$$

Hence

$$\frac{-\zeta'(\sigma)}{\zeta(\sigma)} < \frac{(\sigma-1)^{-2} + \frac{1}{8}\sigma(\sigma+1)^{-1}}{(\sigma-1)^{-1} + \frac{1}{2}} < \frac{1}{\sigma-1} \quad (1 < \sigma \leqslant 5).$$

For $\sigma > 5$ it is enough to check that

$$\frac{-\zeta'(\sigma)}{\zeta(\sigma)} < \sum_{n=2}^\infty \frac{\log n}{n^\sigma} < \frac{\log 2}{2^\sigma} + \int_2^\infty \frac{\log x}{x^\sigma}\,dx < \frac{1}{\sigma-1}.$$

Lemma 70.2. *Let $L(\sigma)$, $X(\sigma)$ be continuously differentiable for $1 < \sigma \leqslant \sigma_0$ and satisfy respectively*

$$-L'(\sigma) \leqslant \varphi(\sigma, L(\sigma)),$$
$$-X'(\sigma) = \varphi(\sigma, X(\sigma)),$$

where $\varphi(\sigma, x)$ is a non-decreasing function of x for each fixed σ. Let $L(\sigma_0) < X(\sigma_0)$. Then $L(\sigma) < X(\sigma)$ throughout the range $1 < \sigma \leqslant \sigma_0$.

Proof. Suppose the contrary. Then there exists σ_1 $(1 < \sigma_1 < \sigma_0)$ such that $L(\sigma_1) = X(\sigma_1)$, $L(\sigma) < X(\sigma)$ for $\sigma_1 < \sigma \leqslant \sigma_0$. And

$$L(\sigma_1) = L(\sigma_0) - \int_{\sigma_1}^{\sigma_0} L'(\sigma)\,d\sigma$$
$$\leqslant L(\sigma_0) + \int_{\sigma_1}^{\sigma_0} \varphi(\sigma, L(\sigma))\,d\sigma$$
$$< X(\sigma_0) + \int_{\sigma_1}^{\sigma_0} \varphi(\sigma, X(\sigma))\,d\sigma = X(\sigma_1)$$

which is a contradiction. This completes the proof.

Proof of Theorem 70. For $y \geqslant 1$, $\sigma > 1$, and integers $q \geqslant 2$, define

$$L(\sigma) = \sum_{m=1}^{\infty}{}^{*} M_q(m)^{1/q} \frac{y^{\omega(m)}}{m^{\sigma}}$$

where the * denotes that summation is restricted to squarefree m. We have

$$-L'(\sigma) = \sum_{m=1}^{\infty}{}^{*} M_q(m)^{1/q} \frac{y^{\omega(m)}}{m^{\sigma}} \sum_{p|m} \log p$$
$$= y \sum_{p} \frac{\log p}{p^{\sigma}} \sum_{\substack{n=1 \\ p\nmid n}}^{\infty} M_q(np)^{1/q} \frac{y^{\omega(n)}}{n^{\sigma}}. \qquad (7.25)$$

We raise (7.23) to the power q and integrate, to obtain

$$M_q(np) = \sum_{\substack{a_0+a_1+\cdots \\ +a_{r-1}=q}} \binom{q}{a_0, a_1, \ldots, a_{r-1}} N(n; \log p; a_0, \ldots, a_{r-1}). \qquad (7.26)$$

There are r diagonal terms on the right (in which one a_i is equal to q), and for these we have $N(n; \log p; a_0, \ldots, a_{r-1}) = M_q(n)$. Lemma 70.1 and the Fundamental Lemma (7.11) yield

$$\sum_{p} M_q(np) \frac{\log p}{p^{\sigma}} \leqslant \frac{rM_q(n)}{\sigma - 1}$$
$$+ (r^q - r)C \cdot 2^{q-1} D_{r-1}(n) \tau_r(n)^{1/(q-1)} M_q(n)^{(q-2)/(q-1)}$$

so that

$$\sum_{p} M_q(np)^{1/q} \frac{\log p}{p^{\sigma}} \leqslant \frac{r^{1/q} M_q(n)^{1/q}}{(\sigma - 1)}$$
$$+ \frac{2rC^{1/q}}{(\sigma-1)^{1-1/q}} D_{r-1}(n)^{1/q} \tau_r(n)^{1/q(q-1)} M_q(n)^{(q-2)/q(q-1)}$$

employing Hölder's inequality, Lemma 70.1, and the inequality $(a+b)^{1/q} \leqslant a^{1/q} + b^{1/q}$. From this and (7.25) we deduce that

$$-L'(\sigma) \leqslant \frac{yr^{1/q}}{(\sigma - 1)} L(\sigma)$$
$$+ \frac{2rC^{1/q} y}{(\sigma-1)^{1-1/q}} \sum_{n=1}^{\infty}{}^{*} D_{r-1}(n)^{1/q} \tau_r(n)^{1/q(q-1)} M_q(n)^{(q-2)/q(q-1)} \frac{y^{\omega(n)}}{n^{\sigma}}, \qquad (7.27)$$

and we apply Hölder's inequality with three exponents $q, q(q-1)$, $(q-1)/(q-2)$ to the sum on the right, which therefore does not exceed

$$\left(\sum_{n=1}^{\infty}{}^{*} D_{r-1}(n) \frac{y^{\omega(n)}}{n^{\sigma}} \right)^{1/q} \left(\sum_{n=1}^{\infty}{}^{*} \tau_r(n) \frac{y^{\omega(n)}}{n^{\sigma}} \right)^{1/q(q-1)} L(\sigma)^{(q-2)/(q-1)}.$$

The occurrence of $D_{r-1}(n)$ here suggests an induction on r. To make this work, we find that q needs to increase with r and we set $q = l + r$, where $l \in \mathbb{Z}^{+}$.

Induction hypothesis. *Let $y \geqslant 1$ be fixed, $l_0(y)$ be sufficiently large. For each $r \geqslant 1$, the proposition $\mathscr{P}(r, y)$ is that for every $l \geqslant l_0(y)$ there exists a constant $C(l, r, y)$ such that, for $1 < \sigma \leqslant 2$,*

$$\sum_{d=1}^{\infty}{}^{*} \Delta_r(d)\frac{y^{\omega(d)}}{d^\sigma} \leqslant \frac{C(l,r,y)}{(\sigma-1)^{ry-r+1+r(r-1)/2(l+r)}}; \tag{7.28}$$

moreover we may take

$$\left.\begin{aligned} C(l,r,1) &= C^{r-1}\cdot 4^r\left\{\prod_{s<r}\left(\frac{56}{s}\right)^s\right\}\left\{\frac{28^{r-1}}{(r-1)!}\right\}^l (l+r)^{(r-1)(l+r-1)},\\ C(l,r,y) &= C^{r-1}\cdot 4^{ry}\left(\frac{10y}{y-1}\right)^{r^2}\left(\frac{5y}{y-1}\right)^{(r-1)l}, \quad (y>1). \end{aligned}\right\} \tag{7.29}$$

The induction hypothesis is that $\mathscr{P}(r-1, y)$ holds.

We notice that $\mathscr{P}(1, y)$ is valid, because the left-hand side of (7.28) is simply $\prod(1 + yp^{-\sigma}) \leqslant 2^y(\sigma-1)^{-y}$ $(1 < \sigma \leqslant 2)$. We have to deduce $\mathscr{P}(r, y)$ from $\mathscr{P}(r-1, y)$.

From the definition (7.10) of D, we have

$$\begin{aligned} \sum_{n=1}^{\infty}{}^{*} D_{r-1}(n)\frac{y^{\omega(n)}}{n^\sigma} &\leqslant \sum_{d=1}^{\infty}{}^{*} \Delta_{r-1}(d)\frac{y^{\omega(d)}}{d^\sigma}\sum_{m=1}^{\infty}{}^{*}\frac{y^{\omega(m)}}{m^\sigma}\\ &\leqslant C(l,r-1,y)2^y/(\sigma-1)^{ry-r+2+(r-1)(r-2)/2(l+r-1)} \end{aligned}$$

and (7.27) becomes (remembering that $q = l + r$)

$$-L'(\sigma) \leqslant \frac{yr^{1/q}}{(\sigma-1)}L(\sigma) + \frac{2rC^{1/q}yC(l,r-1,y)^{1/q}\cdot 2^{y/q+ry/q(q-1)}}{(\sigma-1)^{1+(ry-r+1)/(q-1)+r(r-1)/2q(q-1)}}L(\sigma)^{(q-2)/(q-1)}. \tag{7.30}$$

This differential inequality is the heart of the matter. The corresponding differential equation has solution $X(\sigma) = K(\sigma-1)^{-\gamma}$, where γ satisfies

$$\gamma + 1 = 1 + \frac{ry-r+1}{q-1} + \frac{r(r-1)}{2q(q-1)} + \frac{q-2}{q-1}\gamma$$

or

$$\gamma = ry - r + 1 + r(r-1)/2q,$$

and

$$K = \left(\frac{2ryC^{1/q}C(l,r-1,y)^{1/q}2^{y/q+ry/q(q-1)}}{\gamma - yr^{1/q}}\right)^{q-1}.$$

We want to apply Lemma 70.2 with $\sigma_0 = 2$, and we need $L(2) < X(2) = K$. Now $\gamma \leqslant ry$, $C \geqslant 1$, $C(l, r-1, y) > 4^{(r-1)y}$ from (7.29). For $q > r \geqslant 2$ this implies $K > 2^{ry}$. From (7.7),

$$L(2) \leqslant \sum_{n=1}^{\infty}{}^{*} \tau_r(n)\frac{y^{\omega(n)}}{n^2} < \zeta(2)^{ry} < K.$$

The lemma now yields $L(\sigma) < X(\sigma)$ for $1 < \sigma \leqslant 2$, and together with (7.8) this gives (7.28), with

$$C(l, r, y) \leqslant 2^{r-1}K.$$

We have $r^{1/q} \leqslant 1 + \min\{3r(r-1)/7q,\ 1.4\ \log r/q\}$ for $q > r \geqslant 2$. When $y = 1$, the first bound here yields $ry/(y - ry^{1/q}) \leqslant 14q/(r-1)$ and

$$C(l, r, 1) \leqslant 2^{r-1}\left(\frac{28q}{r-1}\right)^{l+r-1} C \cdot C(l, r-1, 1)\cdot 4$$

so that $\mathscr{P}(r-1, 1) \Rightarrow \mathscr{P}(r, 1)$, with $l_0(1) = 1$. When $y > 1$, we use the second bound for $r^{1/q}$ and we have $2ry/(y - yr^{1/q}) \leqslant 2ry/\{(r-1)(y-1) - 1.4y\log r/q\} \leqslant 5y/(y-1)$ if $7y\log r/q \leqslant (3r-5)(y-1)$: this holds for every $r \geqslant 2$ if $l_0(y)$ is large enough. Thus

$$C(l, r, y) \leqslant 2^{r-1}\left(\frac{5y}{y-1}\right)^{r-1}\left(\frac{5y}{y-1}\right)^{l} C \cdot C(l, r-1, y)\cdot 4^{y}$$

and $\mathscr{P}(r, y)$ again follows. This completes the induction.

Now let $Z_r(x, y)$ be the function defined by (6.12). We put $\sigma = 1 + 1/\log x$ in (7.28) to deduce that

$$\frac{Z_r(x, y)}{(\log x)^{ry-r+1}} \ll_{r,y} \begin{cases} B^l(l+r)^{l+r-1}(\log x)^{r(r-1)/2(l+r)} & (y = 1), \\ (5y/(y-1))^{(r-1)l}(\log x)^{r(r-1)/2(l+r)} & (y > 1), \end{cases}$$

where $B := 28^{27}/27! < e^{30}$. We choose l such that

$$l + r = [(r\log_2 x/\log_3 x)^{1/2}] + 1 \qquad (y = 1),$$
$$l = [(r\log_2 x/2\log(5y/(y-1)))^{1/2}] \quad (y > 1),$$

and apply Theorem 61 to obtain the result stated.

7.4 Second variant – double induction

This version of the method still relies on the Fundamental Lemma, Theorem 73; and this being so the comparison of $M_q(np)$ with $M_q(n)$ remains. We now demand that $p = P^+(np)$ which alters the mechanism by which the extra prime is introduced (requiring another induction over a new parameter k). This suits values of y less than unity, and leads to a proof of Theorem 71.

We need some extra notation. Of course $P^+(n)$ and $P^-(n)$ denote, as usual, the greatest and least prime factors of n, and $p_j(n)$ the jth least. For squarefree n, we set

$$n_k = \begin{cases} \prod_{\leqslant k} p_j(n) & (\omega(n) \geqslant k), \\ n & (\text{else}). \end{cases} \tag{7.31}$$

Lemma 71.1. *Uniformly for $l \in \mathbb{Z}^+$ and $z > 0$, we have*

$$\sum_{\substack{m=1\\ \omega(m)=l}}^{\infty}{}^{*} \frac{1}{m}(\log P^+(m))^{-z} \ll z^{-l}, \tag{7.32}$$

*(where * still means that summation is restricted to squarefree numbers).*

Proof. For fixed p, we have

$$\sum_{\substack{m=1\\ \omega(m)=l\\ P^+(m)=p}}^{\infty}{}^{*} \frac{1}{m} \leqslant \frac{1}{p}\frac{(\log_2 p + c)^{l-1}}{(l-1)!}$$

where c is an absolute constant. Hence the sum on the left of (7.32) is

$$\leqslant \frac{1}{(l-1)!}\sum_p \frac{(\log_2 p + c)^{l-1}}{p(\log p)^z}$$

and the result follows from the Prime Number Theorem and partial summation.

Lemma 71.2. *Uniformly for $0 < y < 1$, $k \in \mathbb{Z}^+$, and squarefree m for which $\omega(m) = k$, we have*

$$\sum_{\substack{n=1\\ n_k=m\\ \omega(n)\geqslant k}}^{\infty}{}^{*} \frac{y^{\omega(n)}}{n^\sigma} \gg \frac{y^k}{m^\sigma}(\sigma-1)^{-y}(\log P^+(m))^{-y}; \tag{7.33}$$

moreover the same sum is

$$\ll \frac{y^k}{m^\sigma}(\sigma-1)^{-y}P^+(m)^{\sigma-1}(\log P^+(m))^{-y}. \tag{7.34}$$

Proof. The integers n on the left take the form $n = mh$, where h is squarefree and either $h = 1$ or $P^-(h) > P^+(m)$. So the sum is

$$\frac{y^{\omega(m)}}{m^\sigma}\prod_{p>P^+(m)}\left(1+\frac{y}{p^\sigma}\right) \asymp \frac{y^k}{m^\sigma}(\sigma-1)^{-y}\exp\{-y\sum_{p\leqslant P^+(m)} p^{-\sigma}\}.$$

By the mean value theorem, there exists $\rho \in (1, \sigma)$ such that

$$\sum_{p\leqslant P^+(m)} p^{-\sigma} = \sum_{p\leqslant P^+(m)} p^{-1} - (\sigma-1)\sum_{p\leqslant P^+(m)} p^{-\rho}\log p,$$

so that the sum on the left on the one hand is $\leqslant \log_2 P^+(m) + O(1)$, which yields (7.33), and on the other hand is $\geqslant \log_2 P^+(m) - (\sigma-1)\log P^+(m) + O(1)$ (since $\rho > 1$), which yields (7.34).

Proof of Theorem 71. For $k, q \in \mathbb{Z}^+$ and $\sigma > 1$ we define

$$L(\sigma; k, q) = \sum_{n=1}^{\infty}{}^{*} M_q(n_k)^{1/q} \frac{y^{\omega(n)}}{n^\sigma}$$

where n_k is defined in (7.31) and * means that $\mu(n) \neq 0$. We consider $L(\sigma; k+1, q)$. If $\omega(n) > k$, we have $n_{k+1} = n_k p_{k+1}$ and we apply (7.26), writing

$$M_q(n_{k+1}) = rM_q(n_k) + E_q(n_k, p_{k+1})$$

where E_q denotes the sum of the off-diagonal terms in which $\max a_i < q$. Thus

$$M_q(n_{k+1})^{1/q} \leqslant r^{1/q} M_q(n_k)^{1/q} + E_q(n_k, p_{k+1})^{1/q}$$

and, if we define $E_q(n_k, p_{k+1}) = 0$ when $\omega(n) \leqslant k$, this holds for all n. Hence

$$L(\sigma; k+1, q) \leqslant r^{1/q} L(\sigma; k, q) + \sum_{\substack{m=1 \\ \omega(m)=k}}^{\infty}{}^{*} \sum_{p > P^+(m)} E_q(m, p)^{1/q} \sum_{\substack{n=1 \\ \omega(n) \geqslant k+1 \\ n_{k+1} = mp}}^{\infty}{}^{*} \frac{y^{\omega(n)}}{n^\sigma}.$$

We estimate the innermost sum on the right by Lemma 71.2, which yields

$$L(\sigma; k+1, q) - r^{1/q} L(\sigma; k, q) \ll \sum_{\substack{m=1 \\ \omega(m)=k}}^{\infty}{}^{*} \frac{y^{k+1}}{m^\sigma} (\sigma - 1)^{-y} \sum_{p > P^+(m)} \frac{E_q(m, p)^{1/q}}{p(\log p)^y}. \tag{7.35}$$

By the Fundamental Lemma (7.11) we have

$$\sum_p E_q(m, p) \frac{\log p}{p} \leqslant r^q C \cdot 2^{q-1} D_{r-1}(m) r^{k/(q-1)} M_q(m)^{(q-2)/(q-1)},$$

and by the Prime Number Theorem, for $z > 0$,

$$\sum_{p > P^+(m)} \frac{1}{p} (\log p)^{-z} \ll \frac{1}{z} (\log P^+(m))^{-z}.$$

We apply this with $z = (qy + 1)/(q - 1)$. Hölder's inequality, with exponents $q, q/(q-1)$, then implies that the inner sum in (7.35) is

$$\ll \frac{2r}{y} C^{1/q} D_{r-1}(m)^{1/q} r^{k/q(q-1)} M_q(m)^{(q-2)/q(q-1)} (\log P^+(m))^{-y-1/q}$$

whence, employing (7.33), we have

$$L(\sigma; k+1, q) - r^{1/q} L(\sigma; k, q)$$

$$\ll \sum_{\substack{m=1 \\ \omega(m)=k}}^{\infty}{}^{*} r D_{r-1}(m)^{1/q} r^{k/q(q-1)} M_q(m)^{(q-2)/q(q-1)} (\log P^+(m))^{-1/q} \sum_{\substack{n=1 \\ n_k = m \\ \omega(n) \geqslant k}}^{\infty}{}^{*} \frac{y^{\omega(n)}}{n^\sigma}$$

$$\ll r \sum_{\substack{n=1 \\ \omega(n) \geqslant k}}^{\infty}{}^{*} \left(\frac{D_{r-1}(n_k)}{\log P^+(n_k)} \right)^{1/q} r^{k/q(q-1)} M_q(n_k)^{(q-2)/q(q-1)} \frac{y^{\omega(n)}}{n^\sigma}. \tag{7.36}$$

We apply Hölder's inequality with exponents q, $(q-1)/(q-2)$, $q(q-1)$ to infer that this is

$$\ll r^{1+k/q(q-1)} R_k(\sigma)^{1/q} L(\sigma; k, q)^{(q-2)/(q-1)} (\sigma - 1)^{-y/q(q-1)}$$

where

$$R_k(\sigma) := \sum_{\substack{n=1 \\ \omega(n) \geqslant k}}^{\infty}{}^{*} \frac{D_{r-1}(n_k)}{\log p_k(n)} \frac{y^{\omega(n)}}{n^\sigma} = \sum_{\substack{m=1 \\ \omega(m)=k}}^{\infty}{}^{*} \frac{D_{r-1}(m)}{\log P^+(m)} \sum_{\substack{n=1 \\ \omega(n) \geqslant k \\ n_k = m}}^{\infty}{}^{*} \frac{y^{\omega(n)}}{n^\sigma}$$

$$= y^k \sum_{\substack{m=1 \\ \omega(m)=k}}^{\infty}{}^{*} \frac{D_{r-1}(m)}{m^\sigma \log P^+(m)} \prod_{p > P^+(m)} \left(1 + \frac{y}{p^\sigma}\right) \ll y^k (\sigma - 1)^{-y}$$

$$\times \sum_{\substack{m=1 \\ \omega(m)=k}}^{\infty}{}^{*} \frac{D_{r-1}(m)}{m^\sigma \log P^+(m)} \prod_{p \leqslant P^+(m)} \left(1 + \frac{y}{p^\sigma}\right)^{-1} \ll y^k (\sigma - 1)^{-y}$$

$$\times \sum_{j=0}^{k} \sum_{\substack{d=1 \\ \omega(d)=j}}^{\infty}{}^{*} \frac{\Delta_{r-1}(d)}{d^\sigma} \prod_{p \leqslant P^+(d)} \left(1 + \frac{y}{p^\sigma}\right)^{-1} \sum_{\substack{h=1 \\ \omega(h)=k-j}}^{\infty}{}^{*} \frac{h^{-\sigma}}{\log P^+(dh)} \tag{7.37}$$

on letting d run through the divisors of m and putting $m = dh$. We put $1 + y = 2z$, so that $z < 1$, and we apply the lower bound $\log P^+(dh) \gg (\log P^+(d))^{1-z} (\log P^+(h))^z$ (where we adopt the temporary convention that $\log P^+(1) = 1$) to the denominator in the innermost sum, and note that, with this convention,

$$(\log P^+(d))^{z-1} \prod_{p \leqslant P^+(d)} \left(1 + \frac{y}{p^\sigma}\right)^{-1} \ll \prod_{p \leqslant P^+(d)} \left(1 + \frac{z}{p^\sigma}\right)^{-1}.$$

By Lemma 71.1 we have

$$\sum_{\substack{h=1 \\ \omega(h)=k-j}}^{\infty}{}^{*} \frac{h^{-\sigma}}{(\log P^+(h))^z} \ll z^{j-k}$$

and, from (7.37) and these last inequalities, we deduce that

$$R_k(\sigma) \ll y^k (\sigma - 1)^{-y} \sum_{j=0}^{k} \sum_{\substack{d=1 \\ \omega(d)=j}}^{\infty}{}^{*} \frac{\Delta_{r-1}(d)}{d^\sigma} z^{j-k} \prod_{p \leqslant P^+(d)} \left(1 + \frac{z}{p^\sigma}\right)^{-1}$$

$$\ll (y/z)^k (\sigma - 1)^{z-y} \sum_{j=0}^{k} \sum_{\substack{d=1 \\ \omega(d)=j}}^{\infty}{}^{*} \Delta_{r-1}(d) \sum_{\substack{n=1 \\ n_j = d \\ \omega(n) \geqslant j}}^{\infty} \frac{z^{\omega(n)}}{n^\sigma}$$

$$\ll (y/z)^k (\sigma - 1)^{z-y} \sum_{j=0}^{k} \sum_{n=1}^{\infty}{}^{*} \Delta_{r-1}(n_j) \frac{z^{\omega(n)}}{n^\sigma} \tag{7.38}$$

employing Lemma 71.2 (7.33) again.

Induction hypothesis. *For each $r \geqslant 1$, the proposition $Q(r)$ is that for every $k \in \mathbb{Z}^+$ and $y \in [0, 1)$ we have*

$$\sum_{n=1}^{\infty} \Delta_r(n_k) \frac{y^{\omega(n)}}{n^\sigma} \ll_{r,y} B(r, y)^{\log^2 k} (\sigma - 1)^{-y}, \tag{7.39}$$

for $1 < \sigma \leqslant 2$, where

$$B(r, y) = (r + 1)^{3/(1-y)}. \tag{7.40}$$

The induction hypothesis is that $Q(r-1)$ holds.

Plainly $Q(1)$ holds: we have to show that $Q(r)$ follows from $Q(r-1)$. We apply $Q(r-1)$ to (7.38) and we obtain

$$R_k(\sigma) \ll_{r,y} k\left(\frac{2y}{y+1}\right)^k B(r-1, z)^{\log^2 k} (\sigma - 1)^{-y},$$

whence, from (7.36),

$$L(\sigma; k+1, q) - r^{1/q} L(\sigma; k, q) \ll_{r,y} k^{1/q} r^{k/q(q-1)} \left(\frac{2y}{y+1}\right)^{k/q}$$
$$\times B(r-1, z)^{\log^2 k/q} L(\sigma; k, q)^{(q-2)/(q-1)} (\sigma - 1)^{-y/(q-1)}. \tag{7.41}$$

This inequality is the analogue, in this variant, of the differential inequality (7.30). We put

$$L_1(\sigma; k, q) := L(\sigma; k, q) + r^{k/q} \sum_{n=1}^{\infty}{}^{*} \frac{y^{\omega(n)}}{n^\sigma}.$$

We see that (7.41) is equally well satisfied by L_1 in place of L. The point of this is that the simple lower bound $L_1(\sigma; k, q) \gg r^{k/q}(\sigma - 1)^{-y}$ is now available, whence (7.41) implies

$$L_1(\sigma; k+1, q) \leqslant L_1(\sigma; k, q)\left\{r^{1/q} + C(r, y) k^{1/q} \left(\frac{2y}{y+1}\right)^{k/q} B(r-1, z)^{\log^2 k/q}\right\}$$

where $C(r, y)$ is suitably large. Now $y < 1$, and so there exists $k_0 = k_0(r, y)$ so large that, for $k \geqslant k_0$,

$$kB(r-1, z)^{\log^2 k} \left(\frac{2y}{y+1}\right)^k \leqslant \exp\{-\tfrac{1}{3}k(1-y)\}.$$

Let $q \leqslant k(1-y)/4 \log k$. We shall then have

$$C(r, y) \exp\left\{-\frac{k}{3q}(1-y)\right\} \leqslant (r+1)^{1/q} - r^{1/q}$$

provided $k \geqslant k_1(r, y)$, and so, for $k \geqslant \max(k_0, k_1)$,

$$L_1(\sigma; k+1, q) \leqslant (r+1)^{1/q} L_1(\sigma; k, q). \tag{7.42}$$

We use (7.42) to derive an upper bound for $L_1(\sigma; k+1, q)$ in terms of $L_1(\sigma; k_2, q)$, where $k_2 = \max(k_0(r, y),\ k_1(r, y))$. Evidently $L_1(\sigma; k_2, q)$. $\ll_{r,y}(\sigma - 1)^{-y}$ because n_{k_2} has at most k_2 prime factors, and $\Delta_r(n_{k_2}) \ll_{r,y} 1$. We want q to be as large as possible, that is $q = [k(1-y)/4 \log k]$, and we have to take account of the fact that q varies with k. Let $q' < q$. From (7.8),

$$M_q(n) \leqslant \Delta_r(n)^{q-q'} M_{q'}(n), \quad \Delta_r(n) \leqslant 2^{r-1} M_{q'}(n)^{1/q'},$$

whence

$$M_q(n)^{1/q} \leqslant 2^{(r-1)(1-q'/q)} M_{q'}(n)^{1/q'}.$$

As we reduce k in (7.42), q decreases occasionally by 1, and we collect a factor

$$\leqslant 2^{(r-1)(1+1/2+1/3+\cdots+1/q)} \leqslant 2^{(r-1)(\log q+1)}$$

altogether, so that

$$L_1(\sigma; k+1, q) \leqslant (r+1)^{\lambda} 2^{(r-1)(\log q+1)} L_1(\sigma; k_2, q_2)$$

where q_2 is the q corresponding to k_2 and

$$\lambda = \sum_{j=k_0}^{k} 1/[j(1-y)/4 \log j] \leqslant \frac{2 \log^2 k}{1-y} + O_y(1).$$

Also $q \leqslant k$, and so $(r-1)(\log q + 1) \leqslant (r-1)(\log k + 1) \leqslant \log^2 k \cdot \log(r+1)$ if (as we assume) $k_2 = k_2(r, y)$ is sufficiently large. Therefore

$$L_1(\sigma; k+1, q) \ll (\sigma - 1)^{-y} \exp\left\{\frac{3 \log^2 k}{1-y} \log(r+1)\right\}$$

and we replace $k+1$ by k, and employ (7.8) to deduce (7.39) and (7.40). This completes the induction from $Q(r-1)$ to $Q(r)$.

We deduce Theorem 71 from (7.39). Write

$$Z_r(x, y) = \sum_{\substack{n \leqslant x \\ \omega(n) \leqslant k}}{}^{*} \Delta_r(n) \frac{y^{\omega(n)}}{n} + \sum_{\substack{n \leqslant x \\ \omega(n) > k}}{}^{*} \Delta_r(n) \frac{y^{\omega(n)}}{n} = Z + Z'.$$

The idea is to make k a large multiple of $\log_2 x$, so that Z' may be estimated trivially. In Z, we have $n = n_k$ and we put $\sigma = 1 + 1/\log x$ in (7.39) which yields

$$Z \ll_{r,y} B(r, y)^{\log^2 k} (\log x)^y. \tag{7.43}$$

Put $k = [er \log_2 x]$. Then

$$Z' \leqslant \sum_{\substack{n \leqslant x \\ \omega(n) > k}}{}^{*} \frac{\tau_r(n)}{n} \leqslant e^{-k} \sum_{n \leqslant x}{}^{*} \frac{(er)^{\omega(n)}}{n} \leqslant \exp\left\{-k + er \sum_{p \leqslant x} \frac{1}{p}\right\} \ll_r 1. \tag{7.44}$$

Since $\log^2 k < \frac{4}{3}(\log_3 x)^2$ for $x > x_0(r)$, (7.3) now follows from (7.43), (7.44) and (6.13).

Notes on Chapter 7

§7.1. The main idea for dealing with the critical interval is due to Tenenbaum (1985) who proved that $\alpha(2,1)=0$. Hall then provided the fundamental lemma and method of differential inequalities, which gave $\alpha(r,1)=0$ $(r \geqslant 3)$.

In the unified development from the fundamental lemma, we introduce the sum

$$S_r^{(t)}(x,y) := \sum_{n \leqslant x} \Delta_r(n)^t y^{\omega(n)} \quad (t \geqslant 1),$$

and set out to prove that

$$S_r^{(t)}(x,y) \ll_\varepsilon x(\log x)^{\beta(r,t,y)-1+\varepsilon} \tag{1}$$

where

$$\beta(r,t,y) := \begin{cases} r^t y - (r-1)t & (y \geqslant (r-1)t/(r^t-1)), \\ y & (y < (r-1)t/(r^t-1)) \end{cases}$$

(see Hall & Tenenbaum (1986) for a more precise result with $(\log x)^\varepsilon$ replaced by a slowly oscillating function of $\log x$). The exponent is sharp, by (6.7). (Incidentally, (1) has not been proved for $t<1$. For $y \leqslant 1$ it is a trivial consequence of (7.2) or (7.3), but for $y>1$ it is an open problem.)

The point is this: in the upper range for y, the differential inequality technique works and yields a result similar to Theorem 70. When $y<(r-1)t/(r^t-1)$ we make the simple observation that, for a suitable $T>t$, we have $y=(r-1)T/(r^T-1)$ and we estimate $S_r^{(T)}(x,y)$ with y now in the upper range, and then apply Hölder.

Exercises on Chapter 7

70. Show that $S_2^{(2)}(x,1) := \sum_{n\leqslant x}\Delta(n)^2 \gg x\log x$.

71. Show that (6.16) and Theorem 70 imply

$$S_2^{(2)}(x,1) \ll_\varepsilon x\log x\cdot\exp\left\{(1+\varepsilon)\surd(2\log_2 x\cdot\log_3 x)\right\}.$$

72. Deduce from Exercise 47 that, for all n,

$$\Delta(n)^2 \leqslant 4\tau(n)^{-1}\sum_{d|n} T(d,0)T(n/d,0)$$

and obtain the upper bound

$$S_2^{(2)}(x,1) \ll x\log x\cdot(\log_2 x)^2.$$

[Hint: Prove an analogue of Theorem 61, and use Theorem 42.]

73. Generalize the results of Exercises 70–72 to

$$S_r^{(2)}\left(x,\frac{2}{r}\right) := \sum_{n\leqslant x}\Delta_r(n)^2(2/r)^{\omega(n)} \quad (r\geqslant 2).$$

74. Show that for r, $n\in\mathbb{Z}^+$ there exists a sub-set of $\mathbb{R}^{r-1}$ of Lebesgue measure $\mu^{(r)}(n)>0$ on which $\Delta_r(n;u_1,u_2,\ldots,u_{r-1})=\Delta_r(n)$. Deduce from this, and (7.7), that $\lim M_q(n)^{1/q}=\Delta_r(n)$.

Appendix: distribution functions

By a distribution function (d.f.) we mean a real valued, non-decreasing function F of the real variable z which is right continuous, i.e. $F(z+0) \equiv F(z)$, and has $F(-\infty)=0$, $F(+\infty)=1$.

Most books on Probability Theory include a section on the main analytic properties of d.f.'s, and we begin with a summary of the most important facts. The proofs are available in, for example, Feller (1966), Loève (1963) or Lukacs (1970).

A d.f. can only have discontinuities of the first kind. A point z is called a *point of discontinuity* if $F(z-0) < F(z)$. The set of discontinuity points is necessarily countable. The complementary set is called the *set of continuity* of F, and is denoted by $\mathscr{C}F$. A *point of increase* of F is such that $F(z+\varepsilon) - F(z-\varepsilon) > 0$ for all $\varepsilon > 0$.

Suppose $\mathbb{R}\backslash\mathscr{C}F$ is arranged in a sequence $\{z_\nu : \nu \geqslant 0\}$ and $s_\nu := F(z_\nu) - F(z_\nu - 0)$ is the *saltus* or *jump* of F at z_ν. Then the function

$$\Phi(z) = \sum_{z_\nu \leqslant z} s_\nu$$

increases only by jumps and is constant in any interval not containing a z_ν. Thus Φ is a step-function and, provided $\mathscr{C}F \neq \mathbb{R}$, it is a constant multiple of a d.f. Such a d.f. is called *purely discrete*, or *atomic*. It is easily verified that $F(z) - \Phi(z)$ is continuous.

A simple example of a continuous d.f. is any function

$$F(z) = \int_{-\infty}^{z} f(t)\,\mathrm{d}t$$

where $f \geqslant 0$ is Lebesgue integrable, and $\|f\|_1 = 1$. Such a d.f. is called *absolutely continuous*. A necessary and sufficient condition that F be absolutely continuous is that

$$\int_N \mathrm{d}F(z) = 0$$

whenever N is a *null set*, i.e. has measure 0. If on the contrary there exists a null set N such that the integral above equals 1, and F is continuous, it is called *purely singular*. The derivative of a purely singular d.f. vanishes almost everywhere.

Theorem A1 (Lebesgue Decomposition Theorem). *Every* d.f. F *can be written, uniquely, in the form*

$$F = \alpha_1 F_1 + \alpha_2 F_2 + \alpha_3 F_3$$

where $\alpha_1, \alpha_2, \alpha_3 \geqslant 0$, $\alpha_1 + \alpha_2 + \alpha_3 = 1$, F_1, F_2, F_3 *are* d.f.'s *such that* F_1 *is absolutely continuous,* F_2 *is purely singular, and* F_3 *is atomic.*

A sequence $\{F_n\}$ of d.f.'s is said to converge weakly to some function F if

$$\lim_{n\to\infty} F_n(z) = F(z) \quad (z \in \mathscr{C}F).$$

Note that the weak limit F is necessarily non-decreasing and bounded, but may not be a d.f.

Let f be a real valued arithmetic function. Then, for every $N \geqslant 1$, the function

$$F_N(z) := N^{-1} \operatorname{card}\{n \leqslant N : f(n) \leqslant z\} \tag{1}$$

is an (atomic) d.f.

Definition. *We say that the arithmetic function f has a limiting distribution (or: has a distribution function) if there exists a* d.f. F *such that the sequence* F_N *defined by* (1) *converges weakly to* F.

We conclude this Appendix with a simple lemma, often used implicitly by Erdős, which provides a convenient sufficient condition for the existence of a limiting distribution for an arithmetic function.

Lemma A2. *Let f be a real valued arithmetic function. Suppose that for each fixed $\varepsilon > 0$ there exists an integer valued function $n \mapsto a(n, \varepsilon)$ such that*

(i) $\lim_{\varepsilon\to 0} \limsup_{T\to\infty} \overline{\mathbf{d}}\{n : n \geqslant 1,\ a(n,\varepsilon) > T\} = 0$,

(ii) $\lim_{\varepsilon\to 0} \overline{\mathbf{d}}\{n : n \geqslant 1,\ |f(n) - f(a(n,\varepsilon))| > \varepsilon\} = 0$,

(iii) *for every a, the asymptotic density* $\mathbf{d}\{n : n \geqslant 1,\ a(n,\varepsilon) = a\}$ *exists.*

Then f has a distribution function.

Proof. For $\eta \to 0+$, fix $\varepsilon = \varepsilon(\eta) \to 0+$ and $T = T(\varepsilon(\eta)) \to \infty$ in such a way that the upper density in (i) is $\leqslant \eta$. Denote the density in (iii) by $\mathbf{d}(a, \varepsilon)$ and put

$$F(z,\eta) := \sum_{\substack{a \leqslant T(\varepsilon) \\ f(a) \leqslant z}} \mathbf{d}(a,\varepsilon), \quad F(z) = \limsup_{\eta\to 0} F(z,\eta).$$

Our first step is to show that F_N, as defined by (1) converges weakly to F. For $z \in \mathscr{C}F$,

$$F_N(z) \leqslant N^{-1} \operatorname{card}\{n \leqslant N: a(n,\varepsilon) \leqslant T(\varepsilon), f(a(n,\varepsilon)) \leqslant z+\varepsilon\}$$
$$+ N^{-1} \operatorname{card}\{n \leqslant N: a(n,\varepsilon) > T(\varepsilon)\}$$
$$+ N^{-1} \operatorname{card}\{n \leqslant N: |f(n) - f(a(n,\varepsilon))| > \varepsilon\}.$$

The first term on the right is $F(z+\varepsilon, \eta) + o(1)$ as $N \to \infty$, by (iii). Taking (i) and (ii) into account, and letting $N \to \infty$ and then $\eta \to 0$, we obtain

$$\limsup_{N\to\infty} F_N(z) \leqslant \limsup_{\eta\to 0} F(z+\varepsilon(\eta), \eta) = F(z).$$

The equality on the right follows from the facts that $F(z', \eta)$ is an increasing function of z' and that z is a continuity point of F. We have, similarly,

$$\liminf_{N\to\infty} F_N(z) \geqslant \limsup_{\eta\to 0} F(z-\varepsilon(\eta), \eta) = F(z).$$

Hence F is the weak limit of F_N, and we may normalize F by making it right continuous. It remains to show that $F(-\infty)=0$, $F(+\infty)=1$. Since $F(z) = \lim F_N(z)$ for $z \in \mathscr{C}F$, evidently $0 \leqslant F \leqslant 1$. Now let $\varepsilon > 0$ and choose $z \in \mathscr{C}F$, $z > \max\{f(a)\colon a \leqslant T(\varepsilon)\} + \varepsilon$. Then $f(n) > z$ implies either $a(n,\varepsilon) > T(\varepsilon)$ or $|f(n) - f(a(n,\varepsilon))| > \varepsilon$. By (i) and (ii), the corresponding density $1 - F(z) \to 0$ as $\eta \to 0$. So $F(+\infty) = 1$. We obtain $F(-\infty) = 0$ in a similar fashion, thereby completing the proof.

Bibliography

M. Balazard

(1987) Remarques sur un théorème de G. Halász et A. Sarközy, preprint.

F.A. Behrend

(1948) Generalization of an inequality of Heilbronn and Rohrbach, *Bull. Amer. Math. Soc.* **54**, 681–4.

A.S. Besicovitch

(1934) On the density of certain sequences, *Math. Annalen* **110**, 336–41.

J.D. Bovey

(1977) On the size of prime factors of integers, *Acta Arith.* **33**, 65–80.

S. Chowla

(1934) On abundant numbers, *J. Indian Math. Soc.* (2) **1**, 41–4.

H. Davenport

(1933) Über numeri abundantes, *Sitzungsbericht Akad. Wiss. Berlin* **27**, 830–7.

H. Davenport & P. Erdős

(1937) On sequences of positive integers, *Acta Arith.* **2**, 147–51.

(1951) On sequences of positive integers, *J. Indian Math. Soc.* **15**, 19–24.

J.-M. Deshouillers, F. Dress & G. Tenenbaum

(1979) Lois de répartition des diviseurs, 1, *Acta Arith.* **34**, 273–85.

Y. Dupain, R.R. Hall & G. Tenenbaum

(1982) Sur l'équirépartition modulo 1 de certaines fonctions de diviseurs, *J. London Math. Soc* (2) **26**, 397–411.

P.D.T.A. Elliott

(1979) *Probabilistic Number Theory* (2 vols.), Springer Verlag (New York).

P. Erdős

(1934) On the density of abundant numbers, *J. London Math. Soc.* **9**, 278–82.

(1935) Note on the sequences of integers none of which are divisible by any other, *J. London Math. Soc.* **10**, 126–8.

(1936) A generalization of a theorem of Besicovitch, *J. London Math. Soc.* **11**, 92–8.

(1946) On the distribution function of additive functions, *Ann. of Math.* **47**, 1–20.

(1948) On the density of some sequences of integers, *Bull. Amer. Math. Soc.* **54**, 685–92.

(1959) Some remarks on prime factors of integers, *Canadian J. Math.* **11**, 161–7.

(1960) On an asymptotic inequality in the theory of numbers (Russian) *Vestnik Leningrad Univ., Serija Mat. Mekh. i Astr.* **13**, 41–9.

(1964) On some applications of probability to analysis and number theory, *J. London Math. Soc.* **39**, 692–6.

(1966) On some properties of prime factors of integers, *Nagoya Math. J.* **27**, 617–23.

(1967) Asymptotische Untersuchen über die Anzahl der Teiler von *n*, *Math. Annalen* **169**, 230–8.

(1969) On the distribution of prime divisors, *Aequationes Math.* **2**, 177–83.

(1979) Some unconventional problems in number theory, *Astérisque* **61**, 73–82.

P. Erdős & R.R. Hall

(1979) The propinquity of divisors, *Bull. London Math. Soc.* **11**, 304–7.

(1980) On the Möbius function, *J. reine angew. Math.* **315**, 121–6.

P. Erdős & I. Kátai
(1971) Non-complete sums of multiplicative functions, *Periodica Mathematica Hungarica* **1**, 209–12.
P. Erdős & J.L. Nicolas
(1976) Méthodes probabilistes et combinatoires en théorie des nombres, *Bull. Sc. Math.* 2° série **100**, 301–20.
P. Erdős & A. Sarközy
(1980) On the number of prime factors of integers, *Acta Sci. Math.* **42**, 237–46.
P. Erdős & G. Tenenbaum
(1981) Sur la structure de la suite des diviseurs d'un entier, *Ann. Inst. Fourier* **31**, 17–37.
(1983) Sur les diviseurs consécutifs d'un entier, *Bull. Soc. Math. de France* **111**, 125–45.
P. Erdős & P. Turán
(1948) On a problem in the theory of uniform distribution I, II, *Indag. Math.* **10**, 370–8, 406–13.
W. Feller
(1966) *An Introduction to Probability Theory and its Applications*, 2 vols., John Wiley, New York.
G. Halász
(1972) Remarks to my paper 'On the distribution of additive and the mean value of multiplicative arithmetic functions', *Acta Math. Acad. Scient. Hung.* **23**, 425–32.
H. Halberstam & H.-E. Richert
(1979) On a result of R.R. Hall, *J. Number Theory* (1) **11**, 76–89.
H. Halberstam & K.F. Roth
(1966) *Sequences*, Oxford University Press.
R.R. Hall
(1974) Halving an estimate obtained from Selberg's upper bound method, *Acta Arith.* **25**, 347–51.
(1974a) The divisors of integers I, *Acta Arith.* **26**, 41–6.
(1975) Sums of imaginary powers of the divisors of integers, *J. London Math. Soc.* (2) **9**, 571–80.
(1978) A new definition of the density of an integer sequence, *J. Austral. Math. Soc.* Ser. A, **26**, 487–500.
(1981) The divisor density of integer sequences, *J. London Math. Soc.* (2) **24**, 41–53.
(1986) Hooley's Δ_r-functions when r is large, *Michigan Math. J.* **33**, 95–104.
R.R. Hall & G. Tenenbaum
(1981) Sur la proximité des diviseurs, *Recent Progress in Analytic Number Theory* (H. Halberstam and C. Hooley eds), Academic Press, London, vol. 1, 103–13.
(1982) On the average and normal orders of Hooley's Δ-function, *J. London Math. Soc.* (2) **25**, 393–406.
(1984) The average orders of Hooley's Δ_r-functions, *Mathematika* **31**, 98–109.
(1986) Les ensembles de multiples et la densité divisorielle, *J. Number Theory* **22**, 308–33.
(1986a) The average orders of Hooley's Δ_r-functions II, *Compositio Math.* **60**, 163–86.
G.H. Hardy & S. Ramanujan
(1917) The normal number of prime factors of a number n, *Quart. J. Math.* **48**, 76–92.
G.H. Hardy & E.M. Wright
(1938) *An Introduction to the Theory of Numbers*, Oxford University Press (fifth edition, 1979).
W. Hengartner & R. Theodorescu
(1973) *Concentration Functions*, Academic Press (London).
A. Hildebrand
(1984) Quantitative mean value theorems for non-negative multiplicative functions I, *J. London Math. Soc.* (2) **30**, 394–406.
(1984a) Fonctions multiplicatives et équations intégrales, *Séminaire de Théorie des Nombres* Paris 1982–83 *Prog. Math.* **51**, 115–24.

(1987) Quantitative mean value theorems for non-negative multiplicative functions II, *Acta Arith.* to appear.

A. Hildebrand & G. Tenenbaum

(1986) On integers free of large prime factors, *Trans. Amer. Math. Soc.* **296**, 265–90.

C. Hooley

(1979) On a new technique and its applications to the theory of numbers, *Proc. London Math. Soc.* (3) **38**, 115–51.

A.E. Ingham

(1930) Notes on Riemann's ζ-function and Dirichlet's L-functions, *J. London Math. Soc.* **5**, 107–12.

I. Kátai

(1976) The distribution of divisors (mod 1), *Acta Math. Acad. Sci. Hungar.* **27**, 149–52.

J. Kubilius

(1964) *Probabilistic methods in the theory of numbers*, translations of Math. Monographs, vol. 11, Amer. Math. Soc., Providence, Rhode Island (third printing, 1978).

L. Kuipers & H. Niederreiter

(1974) *Uniform Distribution of Sequences*, Wiley, New York.

E. Landau

(1909) *Handbuch der Lehre von der Verteilung der Primzahlen*, Teubrer, Leipzig.

P. Lévy

(1937) *Théorie de l'addition des variables aléatoires*, Gauthier-Villars (Paris).

M. Loève

(1963) *Probability Theory*, 2 vols., Springer-Verlag (New York, Heidelberg, Berlin) (fourth edition, 1977).

E. Lukacs

(1970) *Characteristic functions*, Griffin (London) (second edition).

H. Maier

(1987) On the Möbius function, *Trans. Amer. Math. Soc.* **301**, 649–664.

H. Maier & G. Tenenbaum

(1984) On the set of divisors of an integer, *Invent. Math.* **76**, 121–8.

(1985) On the normal concentration of divisors, *J. London Math. Soc.* (2) **31**, 393–400.

E. Manstavičius

(1985) Strong convergence of additive arithmetical functions, *Liteuvos matematikos rinkinys* **25**, no. 2, 127–37 (Russian).

(1986a) The law of iterated logarithm in the Strassen formulation and additive functions, *Lietuvos matematikos rinkinys* **26**, no. 1, 81–90 (Russian).

(1986b) Two laws of the iterated logarithm for additive functions, *Lietuvos matematikos rinkinys* **26**, no. 2, 283–91 (Russian).

J.L. Nicolas

(1984) Sur la distribution des nombres entiers ayant une quantité fixée de facteurs premiers, *Acta. Arith.* **44**, 191–200.

K.K. Norton

(1976) On the number of restricted prime factors of an integer I, *Illinois J. Math.* **20**, 681–705.

(1978) Estimates for partial sums of the exponential series, *J. of Math. and Applications* **63**, 265–96.

(1979) On the number of restricted prime factors of an integer II, *Acta Math.* **143**, 9–38.

(1982) On the number of restricted prime factors of an integer III, *Enseign. Math.*, II Sér., **28**, 31–52.

S. Ramanujan

(1915) Some formulae in the analytic theory of numbers, *Messenger of Math.* **45**, 81–4.

(1915a) Highly composite numbers, *Proc. London Math. Soc.*, Series 2, **14**, 347–409, and *Collected Papers*, Cambridge University Press, 1927, pp. 78–128.

R. Rankin
(1938) The difference between consecutive prime numbers. *J. London Math. Soc.* **13**, 242–7.
I.Z. Ruzsa
(1976) Probabilistic generalization of a number theoretic inequality, *Amer. Math. Monthly* **83**, no. 9, 723–5.
A. Selberg
(1954) Note on a paper by L.G. Sathe, *J. Indian Math. Soc.* **18**, 83–7.
P. Shiu
(1980) A Brun–Titchmarsh Theorem for multiplicative functions, *J. reine angew. Math.* **313**, 161–70.
G. Tenenbaum
(1979) Lois de répartition des diviseurs, 4, *Ann. Inst. Fourier* **29**, 1–15.
(1980) Lois de répartition des diviseurs, 2, *Acta Arith.* **38**, 1–36.
(1982) Sur la densité divisorielle d'une suite d'entiers, *J. Number Theory* **15**, 331–46.
(1984) Sur la probabilité qu'un entier possède un diviseur dans un intervalle donné, *Compositio Math.* **51**, 243–63.
(1985) Sur la concentration moyenne des diviseurs, *Comment. Math. Helvetici* **60**, 411–28.
(1986) Fonctions Δ de Hooley et applications, *Séminaire de Théorie des Nombres*, Paris 1984–85, *Prog. math.* **63**, 225–39.
(1988) Un problème de probabilité conditionnelle en Arithmétique, *Acta Arith.* **49**, to appear.
E.C. Titchmarsh
(1939) *The Theory of Functions*, Oxford University Press (second edition, reprinted 1979).
(1951) *The Theory of the Riemann Zeta-function*, Oxford University Press.
R.C. Vaughan
(1980) An elementary method in prime number theory, *Acta Arith.* **37**, 111–15.
(1985) Sur le problème de Waring pour les cubes, *C.R. Acad. Sci. Paris*, Sér. I, **301**, 253–5.
(1986) On Waring's problem for smaller exponents II, *Mathematika* **33**, 6–22.
(1986a) On Waring's Problem for cubes, *J. reine angew. Math.* **365**, 122–70.
E.T. Whittaker & G.N. Watson
(1927) *A Course of Modern Analysis*, Cambridge University Press (fourth edition).
E. Wirsing
(1967) Das asymptotische Verhalten von Summen über multiplikative Funktionen II, *Acta Math. Acad. Scient. Hungar.* **18**, 411–67.
D.R. Woodall
(1977) A problem on cubes, *Mathematika* **24**, 60–2.

Index

(*Where appropriate we have added key-words to the main entry to guide the reader to the relevant page*)